PLANTS AND HABITATS

An introduction to common plants and their habitats in Britain and Ireland

SECOND EDITION

BEN AVERIS

Published by Ben Averis

www.benandalisonaveris.co.uk

2013

Reprinted with minor changes 2014, 2015, 2018, 2020, 2023

Second edition 2026

ISBN 978-0-9576081-1-5

Printed by Wotton Printers, Newton Abbot, Devon, UK.

The production of the first (2013) print run of this book was supported by financial contributions from Forestry Commission Scotland and Scottish Natural Heritage

Forestry Commission Scotland
Coimisean na Coilltearachd Alba

Scottish Natural Heritage
Dualchas Nàdair na h-Alba
All of nature for all of Scotland
Nàdar air fad airson Alba air fad

Year of Natural Scotland 2013

CONTENTS

ACKNOWLEDGEMENTS

The idea of this book came originally from Andrew Barbour, a landowner and farmer in Perthshire with a very good knowledge of natural history and ecology. He suggested to me that a book showing people how to identify common plants easily and find out about their habitats and relationships with land management would help them 'read' the landscape at all scales and be better placed to appreciate, understand and look after our vegetation. I thank him greatly for encouraging me to write this book, which, I hope, does justice to his vision.

For the first edition many other people gave me suggestions, information and encouragement, and among these I owe special thanks to Alan Crawford (habitat surveyor, Aberdeenshire), Jackie Cumberbirch (Forestry Commission, Aberdeenshire), Dave Genney (Scottish Natural Heritage, Inverness), Steve Hardy, Donald Kennedy (ecologists, in Morvern), Phil Lusby (Royal Botanic Garden, Edinburgh), Stuart Macpherson (East Lothian Council), David Millward (British Geological Survey), my wife Alison (another botanist), my daughter Elen (for advice on layout) and Alison and Elen for putting up with my time working alone on this project. I also acknowledge the references listed on page 385, which I have consulted extensively.

I am very grateful to Forestry Commission Scotland (now Forestry and Land Scotland) and Scottish Natural Heritage (now NatureScot) for making financial contributions toward the cost of the first print run of the first edition of this book, in 2013. Having worked with many FCS/FLS and SNH/NS staff over many years I have great respect for their diligent, hardworking commitment to making progress with knowledge, understanding and conservation.

Since publishing the first edition I have been very encouraged by the number of people who have bought copies of the book, and by plenty of positive feedback. I did not initially expect this book to be so popular for so many years; there have been six print runs of the first edition, totalling 7000 books. I thank everyone who has bought a copy of this book and all those who have given me feedback on it. I hope this second edition is appreciated as well as the first.

2020

INTRODUCTION

Welcome to the second edition of Plants and Habitats, which describes more species than the first edition and is more up to date (see summary box on next page). As in the first edition it focuses mainly on common plant species in Britain and Ireland. The approach is described below under five headings:

1 It's about Plants and Habitats

Most plant books describe species, one at a time, but this book combines information about individual **species** *and* the mixtures of species that together form the **vegetation** in the **habitats** around us. This helps us to understand vegetation at scales ranging from individual plants to whole landscapes, and to see relationships with such things as soils, climate and land management. The 795 species in this book include those that are common, conspicuous or useful as indicators of things such as soil type or particular forms of land management. To help link species with habitats, National Vegetation Classification (NVC) codes are included throughout; this may sound a bit technical but is actually very understandable and for the surveyor brings everything to life even more!

Heather

Bell Heather

Vegetation

2 It's written in plain English

You don't need special botanical knowledge to use this book. You won't be bombarded with difficult technical terms and you won't have to keep flicking back and forth to a glossary. It's all written in plain English. For example, I prefer to describe this leaf (▶) as 'narrow oval, unstalked and hairless, with a pointed tip and untoothed edges' instead of 'ovoid-lanceolate, sessile and glabrous, with an acute apex and entire margins'. Coloured text highlights the main points on identification, habitats and human-related matters such as land management and palatability to grazing animals.

3 It's usable with or without flowers as much as possible

For plant identification, some books rely to a large extent on flowers. That's great when you've got flowers there, but no help when you haven't, so in this book as much identification information as possible is based on longer-lasting things such as leaves and stems. Of course, flowers are described and illustrated too - which happens to make the book a lot nicer to look at! - but wherever possible they are not relied on.

4 Mosses and liverworts are included

They are small but they grow everywhere, make up a large part of many types of vegetation, they help us understand effects of things such as climate and pollution, they're beautiful, and Britain and Ireland is one of the best parts of the world for them. 76 moss and liverwort species are described in this book.

5 The plants are grouped by visual appearance

In most plant books species are grouped into families, which is logical but means that some similar-looking but unrelated species are many pages apart. In this book the ordering is based more on what the plants look like, especially as regards their stems, leaves and growth forms, so that similar but unrelated species are closer together for easier comparison. The ordering is quite simple: it is shown on the Contents page and explained on page 7.

This **SECOND EDITION** is similar to the first edition in structure, layout and size (same number of pages) but with 90 additional species, so it now covers a total of 795 species. Most of the additional species are described quite briefly, with notes about differences from illustrated similar species, but many of them are illustrated too. Some of the original material has been improved, and botanical nomenclature and information sources are updated.

Species included in this book

The species in this book are mostly common, conspicuous or useful as indicators of things such as soil type or particular forms of land management. Collectively they form the bulk of British and Irish vegetation. Some common species are omitted because there is not enough room here for all of them, so I have chosen those that seem likely to be most useful for people to know. Some species that are rather less common are included for their special interest. In Britain and Ireland there are about 5500 plant species growing in the wild. These include over 760 mosses, about 300 liverworts and (for specialists!) more than 970 microspecies of Dandelion (>230), Bramble (>320) and Hawkweed (c.420). Not all of our plants are truly wild and native here. Many are introduced: people brought them here, intentionally or accidentally, from other parts of the world. The most conspicuous species missing from this book are planted crops such as potatoes, though I have included some crop plants: wheat, barley, oats, maize, oil-seed rape and various conifers. Lichens are not true plants and are omitted except for brief notes on seven species (page 337). The book includes the most useful species to know for work with habitat classifications in Britain and Ireland. The choice of species shows a slight northern and western bias because these areas have the wildest landscapes and the largest extents of less modified 'semi-natural' vegetation.

Species descriptions

The text is written in plain English, generally avoiding technical terms, with colour highlighting for main points on **identification**, **habitats** and **human-related matters** (e.g. relationships with land management and palatability to grazing animals), and wherever possible not relying on flowers for identification. Where I have had to use a botanical term, this is explained or illustrated on the same page. The habitats of each species are described in terms of wetness, acidity and broad vegetation type, wherever possible including the main National Vegetation Classification (NVC) community codes. Information about native or non-native status is taken from the Botanical Society of Britain and Ireland (BSBI) *Plant Atlas 2020* (reference given at end of book). Some species have restricted distributions: these are described; other species are widespread in Britain and Ireland. Plant names are those used by the BSBI and the British Bryological Society (website addresses given at end of book). Latin names are written in *italics*. If a page includes repetition of the first word of a Latin name (the *genus*, which is a group of related species) this is shortened to the first letter: e.g. *Sphagnum palustre... S. papillosum...* etc.

Illustrations

The species are illustrated mainly with photographs taken *in situ* to show what they look like in the places where they grow. Most photos include a scale bar to show the size of the plant, but note that size can vary within a species, especially in relation to grazing: plants are generally smaller where more heavily grazed. Not all species are illustrated - there is simply not enough room. However, for species lacking photos there are notes about how they differ from similar-looking species shown on the same page; you need those notes anyway, even with photos. There are also drawings of some species, or of particular features useful for identification.

Habitats section (page 338)

This section describes and illustrates the main habitat types found in Britain and Ireland, to help you interpret habitat names mentioned in the species accounts, to show how different habitat types relate to each other and to environmental factors (climate, soils etc) and land management, and, if only for general interest, to provide an illustrated 'tour' around the habitats and landscapes of our islands. National Vegetation Classification (NVC) codes are included, and at the end of this section are lists of NVC communities and of habitat types from other classifications.

Words used to describe soils and rocks

Throughout this book I have used certain words to describe the types of soils and rocks on which the plants grow. These words, used commonly by botanists, gardeners and many other people, are briefly explained below:

Acid rock is rich in silica and has low amounts of calcium, magnesium and iron. Acid rocks typically give rise to acid soils, which have a pH value of less than 7. Tree bark varies in acidity: for example, that of birches and conifers is more acidic than that of ash and elms.

Basic rock contains less silica and more calcium and magnesium than acid rock. This is reflected in the species composition of the vegetation, but some examples of calcareous grassland and base-rich wetland plant communities, especially in rainy areas where soils are prone to leaching, is on soil with a pH below 7.

Neutral rock and soil is chemically intermediate between acid and basic rock/soil. Neutral grassland plant communities occur on soil whose pH varies from below 7 to just above 7; in many cases, especially but not only where agriculturally improved, the species composition indicates that nutrient richness is higher than is usual in acid and calcareous grasslands.

Mineral soil is composed of material derived from the physical breakdown of rock into small particles and minerals by weathering processes of freezing, thawing, wind and water. It is inorganic: not derived from living organisms.

Peat is an organic surface deposit formed by the dead remains of old plants (especially cottongrasses and *Sphagnum* mosses) that have been compressed by the weight of younger plant material building up above them over thousands of years. It is mostly acidic and can be several metres deep.

Eutrophicated habitats contain high concentrations of nutrients such as nitrates. Eutrophication occurs in various situations: for example, downslope of fields treated intensively with agricultural fertilisers, in water polluted with sewage effluent, and in places with high concentrations of dung or urine from animals such as sheep, cattle and horses. Many plant species are intolerant of excessive eutrophication, but some such as Stinging Nettle are tolerant and can be abundant in eutrophicated habitats.

In this book the species are ordered into three main sections: **Woody**, **Grassy** and **Other**. A summary of each of these groups is given below.

SECTION 1: WOODY plants (page 8). Trees, shrubs, dwarf shrubs and climbers/scramblers. Stems have a tough, woody outer layer.

1a. Broadleaved trees and shrubs (p. 9). Leaves not like needles or scales; mostly falling in autumn, but some are evergreen.

1b. Conifers (p. 45). With needle- or scale-like leaves. Mostly evergreen.

1c. Dwarf Shrubs (p. 50). Shrubs up to about 1 m tall. Includes Thyme and Rockrose: herb-like but woody.

1d. Climbers & Scramblers (p. 722) such as Bramble & Honeysuckle. (Note: five species on p. 78 are not woody.)

SECTION 2: GRASSY plants (p.79); v. narrow leaves; green/brown flower heads.

2a. Grasses (p. 80) *2b. Sedges* (p. 123) *2c. Rushes* (p. 142) *2d. Woodrushes* (p. 146)

SECTION 3: OTHER plants (p. 149) subdivided mainly by leaf shape/arrangement.

3a. Leaf cut into lobes or leaflets along left and right sides (p. 150).

3b. Leaf cut into radiating lobes or leaflets (p. 191).

3c. Whorled (p. 212). Leaves undivided, in groups of 3 or more.

3d. Opposite (p. 221). Leaves undivided, in opposite pairs.

3e. Other herbs (p. 253). Leaves not divided, whorled or opposite. Includes species with leaves only at plant base.

3f. Mosses & other moss-like plants ▸ (liverworts, filmy ferns & clubmosses) (p. 304). Short plants with tiny leaves or just a simple flat, tongue-like 'thallus'. No flowers.

Section 1

WOODY PLANTS: TREES, SHRUBS, DWARF SHRUBS & CLIMBERS/SCRAMBLERS

These plants have stems with a tough, woody outer layer. They are divided here into four groups:

1a. BROADLEAVED TREES & SHRUBS (page 9)

These have leaves of various shapes and sizes, but not like thin needles or tiny scales (except gorse, which has evergreen spines). Most species have leaves falling off in autumn, but some such as holly are evergreen.

1b. CONIFERS (page 45)

These are Firs, Pines, Spruces and so on. Most of them have dark green, needle-like evergreen leaves, but Larch has needle-like leaves that fall off in autumn, and cypresses, Wellingtonia and Western Red Cedar have leaves like little evergreen scales.

1c. DWARF SHRUBS (page 50)

These are shrubs up to about 1 m tall. The most conspicuous and well known are the heathers (which can be useful indicators of the effects of land management); the many other species include thyme and rockrose, which are more herb-like but have woody lower stems.

1d. CLIMBERS & SCRAMBLERS (p.72)

These climb or scramble up, over or among other plants or rocks or ground. Some are palatable to grazing animals and are good indicators of grazing intensity. Five species on page 78 actually have non-woody stems but are described here so as not to split up a distinctive group of hedgerow scramblers and climbers.

Section 1a

BROADLEAVED TREES & SHRUBS

These are trees and shrubs whose leaves are not like thin 'needles' or tiny 'scales' as in Conifers (see page 45), though the Gorses (page 44) are an exception with their thin, sharp-pointed spines. **Shrub** means much the same as **bush** and refers to a woody plant smaller than a tree and in most cases with several main stems instead of one main trunk. Dwarf shrubs - less than about a metre in height - are dealt with separately (page 50). Most broadleaved tree and shrub species are **deciduous** (leaves falling off in autumn) but some such as holly are **evergreen**. For identification it is helpful to divide the broadleaved trees and shrubs into three groups according to the **arrangement of buds (and leaves) along the twigs**:

in **OPPOSITE PAIRS**  page 15

ALTERNATELY to the left and right sides of the twig  page 23

SPIRAL (on all sides of the twig, looking rather random)  page 29

The next five pages give a summary of information for the identification of the broadleaved tree and shrub species in this book: (1) a table listing the species with opposite, alternate and spirally arranged buds/leaves, with each of these three groups sub-divided according to leaf shape; (2) information about the texture and colour of the bark of each species; (3) drawings of typical winter twigs of deciduous species; (4) drawings of typical leaves of these species; (5) photos of some potentially confusing fallen leaves and leaflets in autumn and winter.

Young trees and shrubs are often browsed by animals such as deer, sheep and cattle, so their condition helps us to assess the intensity of grazing. This applies especially to palatable species such as Ash, Rowan, Holly, oaks, willows, poplars, Hawthorn and Blackthorn. Young plants of some species such as oaks, Rowan, Holly, Ash, Sycamore and Beech grow mostly within woodland, while those of others such as birches, Gorse and Broom grow best in the open. Young growth of trees and shrubs is commonly referred to as regeneration.

Bud arrangement and leaf shape. This table groups the deciduous broadleaved trees and shrubs described in this book by a combination of bud arrangement and leaf shape.

BUDS	LEAF SHAPE		SPECIES
OPPOSITE 	Oval		**Dogwoods, Spindle, Snowberry, Fuchsia, Wayfaring-tree, Butterfly-bush** (+**Privet, Box** and **Wilson's Honeysuckle**, all of which are evergreen)
	Cut into lobes in **a spreading, radiating pattern**		**Sycamore, Field Maple, Norway Maple, Guelder Rose**
	Leaf **divided into many leaflets**	Leaflets **in pairs** (L & R) + one at end	**Ash** (9-15 leaflets) **Elder** (3-11)
		Leaflets spread out in **radiating** pattern	**Horse Chestnut**
ALTERNATE 	Oval	with the edges smooth **(no teeth)**	**Beech**
		with **little teeth** along the edges	**Elms, Hazel, Hornbeam** (+ **Cherry Laurel, Portugal Laurel** & **Cotoneasters**, all of which are evergreen)
	Triangular		**Birches**
	Heart-shaped		**Limes**
	Cut into lobes in **a spreading, radiating pattern**		**London Plane**
SPIRAL 	Oval	with the edges smooth **(no teeth)**	**Rhododendron** (evergreen), **Broom, Sea-buckthorn**, some **Willows**
		with **little teeth** along the edges	**Alders, Cherries, Apples, Pears, Plum, Blackthorn, Sweet Chestnut**, most species of **Willow**
		with bigger **lobes** along the edges	**Oaks, Holly** (evergreen), some leaves of **Hawthorns**
	Triangular	**without lobes**	most species of **Poplar**
		with lobes	**Gooseberry** some leaves of **Hawthorns**, some leaves of **White Poplar**
	Round (+ wavy edges + long, flattened stalk)		**Aspen** (a poplar)
	Leaf **divided into many leaflets**	Leaflets **in pairs** (L & R) + one at end	**Rowan, Walnut**
		1-3 leaflets (very small leaf)	**Broom**
	Narrow spines or small scales		**Gorse** (evergreen)

Bark. This table describes the typical bark of the deciduous broadleaved trees and shrubs in this book. Note that young specimens of all species tend to have smoother bark.

SPECIES	BARK TEXTURE **S** = smooth; **M** = medium; **R** = rough		BARK COLOUR Light; Medium; Dark	
Ash	M	finely to rather deeply furrowed	L–M	grey
Walnut	R	Deeply furrowed	L	pale greyish
Elder	R	deeply furrowed	L–M	pale brown
Sycamore	M	flaky in old trees	M	greyish
Norway Maple	M	finely furrowed	M	grey-brown
Field Maple	M	rather scaly	M	brown
Horse Chestnut	M	rather flaky in old trees	M–D	brown
Guelder Rose, Wayfaring-tree, Dogwoods, Privet, Spindle, Snowberry, Fuchsia & Butterfly-bush	S	smooth (Spindle stem = square-sectioned)	L–M	greyish; twigs of Dogwoods purple, Fuchsia red, Spindle green
Birches	M–R	smooth (white patches + horizontal dark lines) + some cracks. In Silver Birch, old lower trunk grey & deeply furrowed.	L	grey & white
Hazel	S	smooth to fairly smooth	M	brown
Beech	S	smooth	L–M	grey
Elms	R	deeply furrowed in old trees	M	grey-brown
Limes	M	smoothish or finely furrowed	M	brownish
London Plane	M	flaky	L/M/D	greyish
Hornbeam	S–M	can have wide ridges/hollows	M	grey
Oaks	M–R	rather deeply furrowed	M	greyish
Rowan & Whitebeams	S	smooth - slightly rough	L–M	grey
Alders	M–R	furrowed	D	dark brown
Grey & Eared Willows	S	older bark only slightly rough	L–M	grey
Goat Willow	S–M	can be deeply furrowed	L–M	grey-brown
Bay Willow	M	finely furrowed	M	grey-brown
Crack/White Willows	R	deeply furrowed	M	brown
Osier + Purple Willow	S	smooth or smoothish	M	greyish
Aspen + White/Grey Poplar	S–M	some horizontal lines/furrows	L–M	greyish
Hybrid Black-poplar	M–R	deeply furrowed on old trunks	M–D	grey-brown
Wild cherry	S–M	smooth, shiny + horizontal lines; old bark furrowed	M–D	brown
Bird cherry & Plum	S–M	smoothish	M–D	grey-brown
Apple	S–M	can get slightly scaly or furrowed	M	grey-brown
Pear	R	divided into many small squares	M	grey-brown
Sweet Chestnut	R	old trunks v. furrowed & gnarled	M	grey-brown
Hawthorns	M	can be scaly or furrowed	M–D	greyish
Blackthorn	S	finely furrowed	D	grey-black
Sea-buckthorn	S	rather scaly	M	greyish
Broom	R	twigs strongly ridged	M	green

Winter twigs. Typical winter twigs of the deciduous broadleaved trees and shrubs described in this book are illustrated below, except for Hawthorns, Blackthorn, Sea-buckthorn and Gooseberry (all thorny) and Broom (green and strongly longitudinally ridged). The species are grouped according to the arrangement of buds along the twigs (opposite, alternate or spiral).

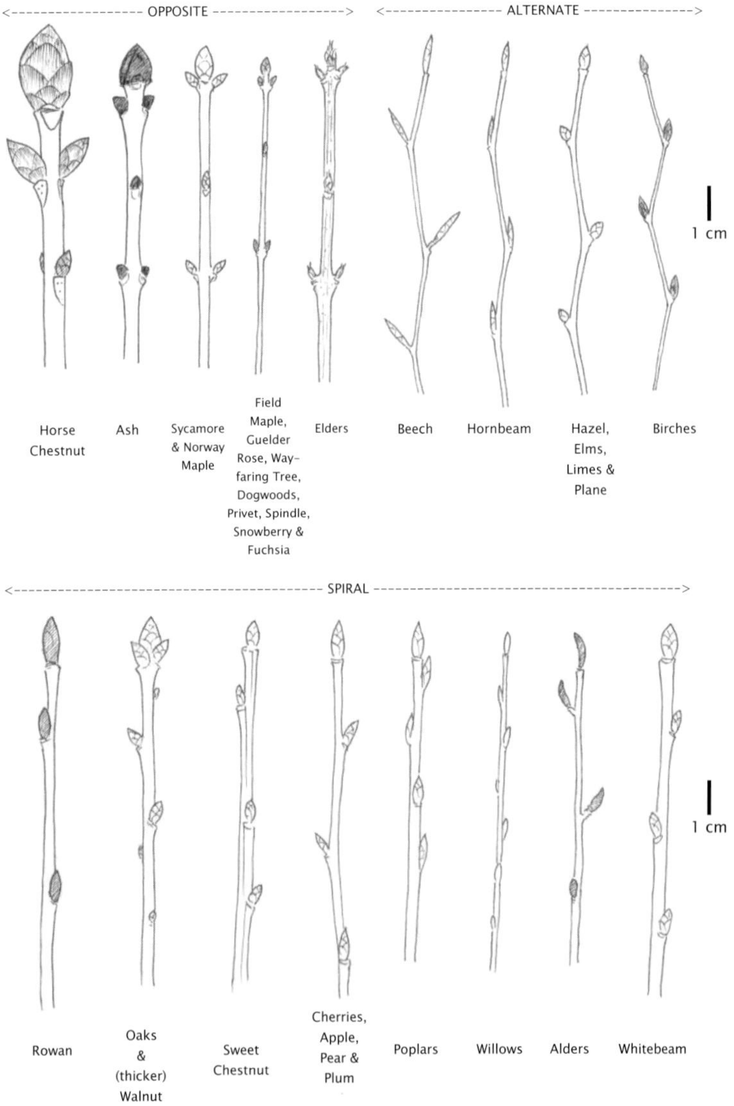

Leaf shape. Typical leaves of broadleaved trees/shrubs in this book are illustrated below, except for:

Bay & Purple Willows (rather like Grey/Goat Willows), Gooseberry (like Field Maple but smaller and more toothed), Box (like Privet but smaller and blunter-tipped), Cotoneasters and Wilson's Honeysuckle (small, oval, pointed, v. short-stalked & evergreen); Walnut (like Ash/Elder but leaflet edges untoothed) and Italian Alder (like Lime but with symmetrical base and less finely-pointed tip).

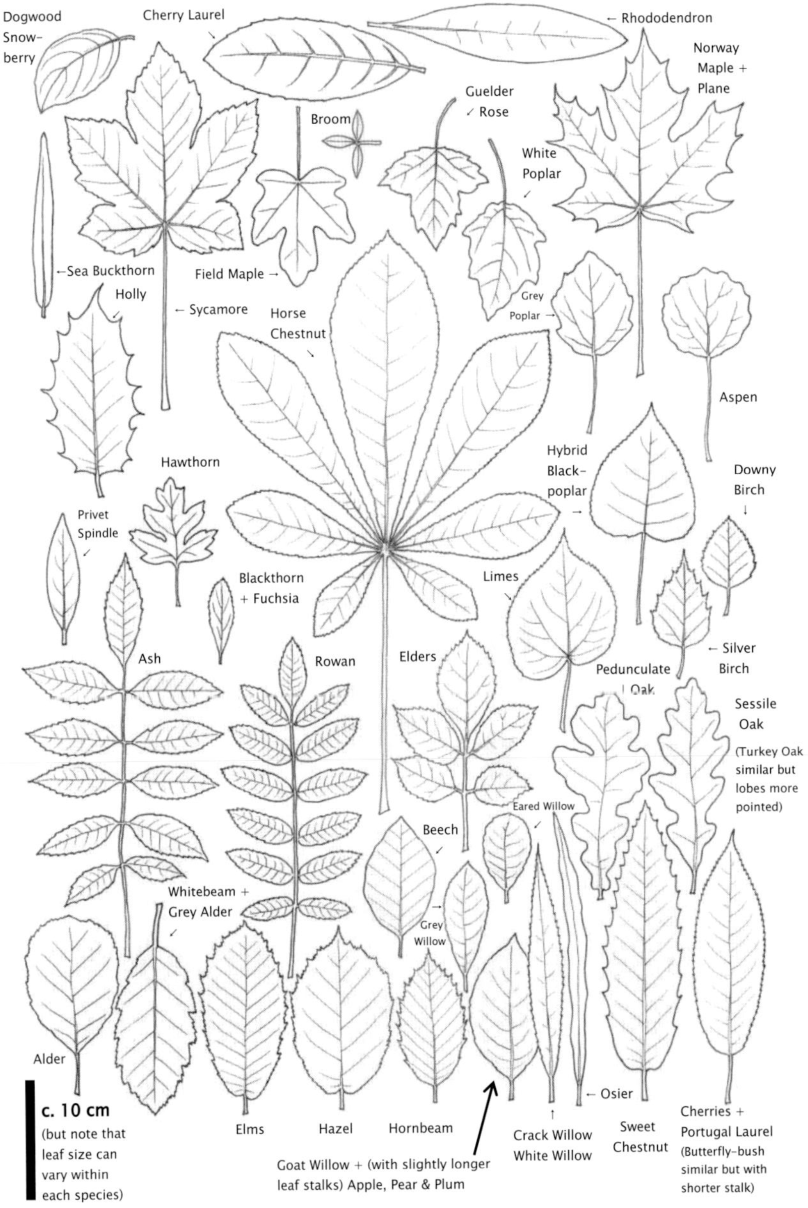

Fallen leaflets of Rowan, Ash & Elder in autumn/winter compared with willow leaves

In autumn and winter fallen leaves on the ground can help with identification. Those of Ash (page15), Elder (page16) and Rowan (page 30) are divided into many separate leaflets that tend to break off the main leaf stalks so they look like whole undivided leaves (especially Willow leaves) with no obvious indication of having been parts of larger 'compound' leaves, though the fallen long, bare leaf stalks give a clue. **Rowan** leaflets are easy to tell: oblong-oval with shortly pointed tips and rounded bases, teeth along the edges from the middle of the leaf to the tip, and with no stalk (Willow leaves have stalks). They commonly turn red in autumn (Ash, Elder and Willows do not). **Ash** leaflets taper more gradually at the tip and base, they have very short 'false stalks' where the base of the leaf blade is very narrow and folded up on each side to leave a groove running along the middle (Willow leaf stalks are more solid and tough and have no groove) and have a more flimsy texture than most Willow leaves. **Elder** leaflets are larger and of a more oval (less oblong) shape, like leaves of Goat Willow (page32) but with sharper teeth, a softer, more flimsy texture and again with 'false stalks'. On the undersides of leaflets the side veins are more prominent and raised in Elder than in Ash and Rowan.

Ash *Fraxinus excelsior*

Leaf

Ash has **thick pale grey twigs** with **thick black buds**, and **large leaves with 9-15 leaflets**. **Bunches of small purplish flowers** in spring (before leaves emerge), become **'keys'** in autumn (green at first, then brown). Leaves are like those of Rowan (page30) but in opposite pairs and the leaflets taper more gradually at their tips and bases. Autumn leaves turn yellow. When they fall many leaflets come off the leaf stalks; on the ground we then see a mix of bare leaf stalks and single leaflets looking like willow leaves (see page 14). Ash grows mainly on **moist to dry, neutral to basic soils** (in woods mainly in W7-9). **Regeneration** mainly in **shade** of other trees; can be fast-growing. The future of Ash in Britain is uncertain now that the fungal disease Chalara (ash dieback) is here.

Purplish flowers in April

Yellowing leaves in autumn

Tree in winter

Winter twigs, with old brown keys

Elder *Sambucus nigra*

Buds in autumn

A **shrub** with **green or brown twigs, green or red-purple buds**, and **leaves with 5-7 oval, toothed leaflets** turning yellowish in autumn. **Wide, flat-topped clusters of flowers and edible fruits**. On **dry to moist, neutral soils** in **woods** (W6-11), **scrub** (W21-23) & **hedges**. Bark can be a good habitat for mosses, liverworts, lichens & fungi.

Leaf in May

Creamy-white flowers in summer

Older bark is rough, pale brown to yellowish-grey, and cork-textured

Black berries in autumn

Elder in June

Red-berried Elder (May)

RBE (July)

RBE buds

Red-berried Elder *S. racemosa* has **smoother, pinker-brown bark**, **fatter, blunter buds** and **narrower, elongated clusters of flowers** (spring) & **red berries** (summer). **Introduced**. On **dry or moist, neutral to acid soil in woods** (W6-11) & **scrub** (W21), esp. in **E Scotland**.

Sycamore *Acer pseudoplatanus*

A tree with **thickish twigs, yellowish-green buds** and **large, long-stalked leaves with radiating lobes** that are toothed and pointed. Hanging spikes of yellow-green flowers become **two-winged 'keys'** by late summer. Leaves mottled green/brown/yellow/blackish in autumn. **Introduced** and **naturalised**, mainly on **dry to moist, basic to slightly acid soils**; in woods mainly in W8–10.

Leaves and flowers in May

Winter twig

Old sycamore (L) and oak (R)

Leaves in May

'Keys' in October

Bark can be flaky

Norway Maple *A. platanoides*: similar but has **brown buds** and **leaves with fewer but larger and more sharply-pointed teeth**. Autumn leaves brighter yellow-orange. Yellowish flowers in round clusters before leaves open. Mainly **planted** and in the lowlands. Plane (page 26) has similar leaves but alternate buds and flaky bark.

Field Maple *A. campestre* is a **smaller** tree with **thinner twigs, brown buds** and **smaller leaves with rounded lobes**. Leaves turn yellow in autumn. Bark browner than in Sycamore. Grows on **dry, neutral to basic soils**; in woods (mainly W8) and hedges; **native in England & Wales**; **planted** in Scotland & Ireland.

Norway Maple leaf in October

Field Maple in May

Horse Chestnut *Aesculus hippocastanum*

This unmistakeable tree has **very thick twigs** with **large sticky brown buds**, **big, long-stalked leaves with 5-7 leaflets in a radiating pattern**, **large spikes of white flowers** in spring, and **round, green, spiky fruits** containing shiny brown **'conkers'** in autumn. Leaves turn yellow-orange in autumn. Bark brown; can be flaky in older trees. **Planted**, mainly in the lowlands.

Leaves and flowers in May

'Sticky buds' in early spring

'Conkers' in September

Guelder Rose *Viburnum opulus*

A **shrub** with fairly slender twigs (older ones greyish), small oval, pointed buds, and **stalked leaves with two large side lobes** (and some smaller teeth on the lobe edges). Leaves turn pink-red in autumn. Clusters of **white flowers** in summer; **red berries** in autumn. Grows mainly on **damp neutral to basic soils**, in **woodland** (esp. W7-9), **scrub** (W21) and **hedgerows**. Commonest in the **lowlands**, especially in England and Wales.

Shoots in June (above) and October (below)

Wayfaring-tree *Viburnum lantana*

A **shrub** with rather thick, pale greyish twigs, **oval leaves with toothed edges and a coating of short hairs on both sides** (especially the undersides). The small **whitish flowers** appear in spring and are in dense clusters. They develop through summer into **berries** that are longer than wide and change colour from green to red and finally black.

Shoot in June, with very young berries

On **dry, basic soils** in **woods, scrub** and along **hedgerows**. **Native** in the lowlands S of a line from Plymouth through Swansea, Hereford and Lincoln to Ipswich. **Planted** in many places further N and W, but rare in Ireland.

Dogwood *Cornus sanguinea*

A **shrub** with slender **twigs (purplish in winter)** with small oval, pointed buds, and **oval, untoothed, short-stalked leaves** whose **veins (4 pairs on each leaf) curve so they do not reach the leaf margin**. Leaves turn dull purplish in autumn. Clusters of **white flowers** in summer; **purple-black berries** in autumn. On **dry basic soils** in **southern British lowlands**, (native north to S Cumbria & S Northumberland) in **woods** (esp. W8), **scrub** (esp. W21) and **hedges**.

Shoots in June (above) and October (below)

Red Osier Dogwood *C. sericea* has **5-7 pairs of veins per leaf**. **Introduced**. By **roads** and **rivers**, in **parks** and **woods**; mostly in lowlands.

Spindle *Euonymus europaeus*

A **shrub** with fairly slender, **square-sectioned twigs** (young ones green; older ones browner), small oval pointed buds and **stalked, oval, pointed leaves with many small teeth** along their edges. Leaves turn pinkish in autumn. Small clusters of small flowers in summer (each with 4 narrow pale green petals); **strange-shaped pink berries** in autumn. Grows mainly on **dry neutral to basic soils**, in **woodland** (esp. W8), **scrub** (W21) and **hedgerows**. Mainly in the **lowlands**. Rare in Scotland.

Privet *Ligustrum vulgare*

A **shrub** with **evergreen or semi-evergreen, oval, untoothed, shortly-stalked leaves** up to 6 cm long, on **slender, very shortly hairy twigs**. Spikes of **small white flowers** in summer; **black berries** in autumn. On **dry, neutral to basic soils** in **lowland woods** (esp. W8 & W10) and **hedges**. **Native** in S half of Britain. **Introduced** in Scotland, Ireland & Northumberland. **Commonly planted as pheasant cover.**

Privet berries

Garden Privet *L. ovalifolium* is similar but has **hairless twigs** and **wider and more consistently evergreen leaves**. Commonly **planted in hedges**. Naturalised on **disturbed/waste ground**. Mainly near human habitation.

Box *Buxus sempervirens* ▸ is a shrub with **small oval, blunt-ended, shiny/leathery evergreen leaves** <3 cm long. Widely **planted** in lowland **woods/hedges**. Very rare native in woods in SE England.

Wilson's Honeysuckle *Lonicera nitida* is like Box but **leaves smaller (<1.5 cm long), widest below middle** and **less thick/leathery**. **Planted**, mainly in urban environments.

Box leaves

Snowberry *Symphoricarpos albus*

A **shrub** with slender **twigs (purplish in winter)**, small oval, pointed buds, and **untoothed oval leaves with blunt or shortly-pointed tips and short stalks. Small pink flowers** (summer); **round white berries** (autumn, persisting through winter). **Introduced**. Mainly on **neutral to basic soils** in **lowland woods** (esp. W7/8/10), **scrub** (W21) and **hedges**.

Fuchsia *Fuchsia magellanica*

A shrub 1-3 m tall, with slender red twigs, dark green or purple-green oval, toothed leaves on short red leaf stalks, and hanging pink-red flowers in summer/autumn. Introduced. Mainly in the west (esp. Ireland), in hedges, along walls, by streams and on rocky ground.

Butterfly-bush *Buddleja davidii*

A shrub 1-3 m tall, with rather thick brownish twigs. Oval, pointed leaves up to 25 cm long, with finely toothed edges, dark green uppersides and whitish, felt-textured undersides. Small leaf-like stipules at bases of short leaf stalks. Conspicuous long, tapering spikes of small pale purple flowers at branch tips in late summer; green-brown in fruit in autumn. Flowers much frequented by butterflies. Introduced: on dry soils in gardens, towns & villages, on disturbed ground, on walls and by railway lines.

Birches *Betula* species

Graceful trees with very thin, delicate twigs bearing oval buds and small, triangular or diamond-shaped leaves with toothed edges. Leaves yellow or orange in autumn. Bark has **white patches** but old lower trunks, especially on Silver Birch, can be grey and rugged. Flowers/fruits in small catkins. **Mainly on acid soils**. **Cold-tolerant**; very common in Scottish Highlands. Can colonize open ground, forming thickets. Regeneration **moderately palatable** to deer, sheep & cattle. See also Dwarf Birch (shrub with round leaves) on page 65.

Downy Birch *Betula pubescens* ▸ has **short hairs on the twigs, its buds are blunt** and its **leaves have single-toothed edges** (teeth all about the same size) and **tips not drawn into long, fine points**. On **dryish to wet acid** (less commonly neutral to basic) **soils**; in woods mainly W2-7, 9-11 & 16-18.

Silver Birch leaves and catkins

◂ **Silver Birch** *B. pendula* has **hairless twigs** (with tiny warts), **pointed buds**, and leaves with **double-toothed edges** (larger teeth at intervals among smaller ones) and **tips drawn into long, fine points. Old trunk grey, with deep, more or less vertical furrows**. On **dry to moist, acid soils**; in woods mainly in W10-11 & W16-18.

Hazel *Corylus avellana*

A **multi-stemmed shrub** with brown twigs (young ones hairy), **short, blunt, green buds** and **broad oval leaves** with slightly heart-shaped bases and **double-toothed edges** (many small teeth and fewer larger teeth). Leaves turn yellowish in autumn. **Yellow catkins** in spring. **Edible nuts** in autumn. Grows on **dry to moist, neutral to basic soils**; in woods mainly in W8-11. **Coppiced** in places, especially in the past, but hazels naturally 'self-coppice': old, thick stems lean out and fall off; young thin stems grow vertically from middle (but can be bitten off by sheep/deer). In W Scotland hazels (especially old uncoppiced ones) can support **rich moss, liverwort & lichen** floras.

Western Scottish hazel woodland. Inset: hazel nuts in October.

Leaves turning yellow in October

Catkins in February

Elms *Ulmus* species

These trees have **brown twigs**, **short brown buds** and **oval, rough-textured leaves** with **double-toothed** edges (larger teeth at intervals among more numerous smaller ones) and **asymmetrical bases**. The leaf stalks are very short. Leaves turn yellow in autumn. There are **small tufts of purple flowers** in early spring (out before the leaves), soon becoming **small green flattish fruits**. Elms grow on **dry to moist, neutral to basic soils**.

Wych Elm *Ulmus glabra* has **long oval leaves 8-16 cm long** and its **older twigs lack cork-like ridges**. Grows in **woods** (mainly W8-9) and also in **hedges** and on **cliffs**. Native throughout Britain and Ireland.

English Elm *U. procera* has **smaller, rounder leaves <9 cm long**, has **cork-like ridges on older twigs** and grows mainly in **lowland hedges**. It can be a tall tree; most of these trees have been killed by Dutch Elm disease but the species is widespread in **lowland hedges**. Possibly native in S half of Britain.

Small-leaved Elm *U. minor* is like English Elm but its leaves have **smooth, hairless uppersides** (rough & hairy in the other elms). Mainly in **hedges in English lowlands**.

Wych Elm leaves

Wych Elm twig

Wych Elm flowers (April)

English Elm leaves

Wych Elm fruits (May)

Limes *Tilia* species

Trees with **reddish zig-zag twigs, oval, shortly-pointed red-brown buds,** and **long-stalked heart-shaped leaves with unmistakeably asymmetrical bases.** The fruits are also distinctive (see photo). Leaves yellowish in autumn. **Young leaves are edible**. Many trees have a dense growth of short, young shoots around the trunk in the lower half of the tree (see upper-left photo below).

Common Lime *T.* x *europaea*, our commonest Lime, is a hybrid between the next two species. Its leaves are **6–10 cm long** and **hairless except for small tufts of hairs at the junctions of veins on the underside**. **Planted**, mainly in lowlands and especially near human habitation (parks, avenues etc).

Small-leaved Lime *T. cordata* is similar but has **smaller, rounder leaves only 3–6 cm long**. It is native and grows mainly on **neutral soils in woods** (W8-11) in England and Wales.

Large-leaved Lime *T. platyphyllos* has **larger leaves 6–12 cm long and hairy on both sides**. As a native species it has a widespread but patchy distribution in England and Wales, growing in **woods** and on **cliffs**, but it is also **planted** elsewhere.

Common Lime

Common Lime

Common Lime leaf

Common Lime leaves and fruits

London Plane ► *Platanus x hispanica* has rather similar twigs but has **flaky bark** patterned light & dark, **lobed leaves** like those of Norway Maple (p. 17), and **spherical fruits on hanging stalks**. **Planted** in many **towns & cities**.

Beech *Fagus sylvatica*

A large tree with **smooth grey bark, long (1–2 cm) thin pointed shiny brown buds sticking out from the twig, and oval untoothed leaves** that are bright yellow-green (and **edible**) in spring, mid green in summer, turning orange-brown in autumn; can remain through winter on beech hedges. On **dry acid to basic soils**; in woods mainly W12 & 14–15. Regeneration mainly in shade of other trees. **Native in SE Eng. & SE Wales; introduced & naturalised elsewhere.**

Leaves in June

Fallen leaves in September

Twig in April

Beech mast in September

Tree in early spring

Trees in late autumn

Hornbeam *Carpinus betulus*

This tree has **narrow pointed brown buds** as in beech, but **shorter (<1 cm) and pressed against the twigs**. The **oval, pointed leaves with double-toothed edges** (larger teeth at intervals among smaller ones) are like those of elm (but more or less symmetrical at the base, and not hairy) and hazel (but smaller, narrower and hairless). Bark grey and quite smooth, older trunks with furrows running up at an angle as if the trunk has been twisted. **Native** in S England in lowland **woods** (esp. W10) **on neutral soils**; **planted** trees more widespread.

Leaves in May

Leaves and fruits in June

Cherry Laurel *Prunus laurocerasus*

A **large-leaved evergreen shrub** like Rhododendron (page 42) but leaves **alternate**, **blunter-tipped**, with **small teeth** and **prominent pale veins**; side veins curl forwards. **Leaf stalk short** (<15 mm). Spikes of **small white flowers** in spring, later becoming **blackish berries**. **Introduced**; mainly **planted** in lowland woods/gardens. Generally **less invasive** than Rhododendron.

Cherry Laurel: leaves in 2 rows along stems

Rhododendron: leaves stick out in all directions

Portugal Laurel *P. lusitanica*: rather like Cherry Laurel but **leaves smaller**, with **more pointed tips** and **longer (>15 mm) reddish stalks**; leaf shape superficially similar to that of the non-evergreen Bay Willow. **Planted** in lowlands, esp. in gardens and urban areas.

Cotoneasters *Cotoneaster* species: shrubs with **small (mostly <2 cm long), v. short-stalked, oval, pointed evergreen leaves**, small **white flowers** & **red berries**. Many species. **Planted** in gardens, etc. Can spread into wild, especially in low altitude open or rocky habitats.

Oaks *Quercus* species

Familiar trees with distinctive leaves: **oblong, with rounded lobes and short stalks**, turning brown or orange-brown in autumn. Bark greyish; older trunks furrowed. Fairly thick twigs have brown **buds rather clustered at tips**. Yellow-green catkin-like flowers in spring. Unmistakeable **acorns** in autumn. Common on **dry to moist, neutral to acid soils**; in woods mainly in W10-11 & 16-17. Regeneration is slow and mainly in shade of other trees. Habitats and distributions of our two native species (**Pedunculate** & **Sessile** Oaks) overlap.

Above: Pedunculate Oaks in lowland woodland. Right: twig in April and acorns in September

Pedunculate Oak *Q. robur*: **leaf base 'abrupt' with two small lobes; leaf stalk <1 cm long**; **acorns on long stalks**; <20 scales on outer surface of buds; commonest in lowlands, but also occurs in upland areas.

Sessile Oak *Q. petraea*: **leaf base tapers gradually into longer (1 cm) stalk**; **acorns unstalked**; leaf lobes more evenly sized and spaced; >20 scales on outer surface of larger buds; commonest in upland areas but also present in lowlands.

NOTE: **Pedunculate-Sessile hybrids** are common and have leaves of **intermediate** shape with slightly lobed bases or with more distinctly lobed bases and leaf stalks >1 cm long.

Three **introduced** species in **lowlands**: **Turkey Oak** *Q. cerris* (commonest of the 3) has **narrow leaves** with **rather pointed lobes**, & **long bristles on buds**. **Red Oak** *Q. rubra* has **large leaves** (**red** in autumn), with **fewer, larger, pointed lobes**. **Evergreen Oak** *Q. ilex* (see p. 38) has **dark, leathery, evergreen leaves**.

Rowan *Sorbus aucuparia*

Also called **Mountain Ash**. A small tree with smoothish grey bark and **rather thick greyish twigs with large grey-brown buds. Leaves with 7–9 pairs of leaflets**, turning orange-red in autumn. Leaves like Ash but not in opposite pairs; leaflets more abruptly-tapered at bases and tips. Clusters of **creamy-white flowers** in spring; **red berries** (good for **edible** jelly) in autumn. On **dry to moist, acid to basic soils**; in woods mainly in W9/11/17. Regeneration mainly in shade of other trees, and faster-growing than that of oak. **Bark and leaves palatable to sheep, deer, etc**.

Leaves and flowers in May

Leaves and berries in August

Young tree in April

Fallen leaflet in autumn (rounded base and no stalk; reddish colour)

1 cm

Fallen Ash leaflet (tapering base and short stalk; no red)

Walnut leaf in late summer

Walnut *Juglans regia*: a tree with ash-like leaves but **leaflets larger (to >20 cm long), fewer (2–4 pairs), untoothed**. Thick twigs. Pale bark. Male flowers: hanging catkins. Female: shorter, becoming spherical fruits + **edible nuts**. **Planted/naturalised** in **lowlands**.

Whitebeams *Sorbus* species ▸ are like Rowan but **leaves are oval, with toothed or lobed edges and whitish undersides**. Many species (can be hard to separate): on **neutral to basic soils** in **woods** and (esp. in W) on **cliffs**; **also planted**.

Alder *Alnus glutinosa*

Leaves are **broad oval, toothed dull green, widest around or beyond the middle**, turning dull, dark green-brown in autumn. **Catkins** small and green in winter; long, yellow and hanging in spring. **Cone-like fruits** are green at first, then blackish, remaining on the tree for at least a year. On **moist to wet, neutral soils**; in woods mainly in W3-7. Habitats include **stream/river banks** and **edges of ponds/lakes**.

Alder woodland in August

Young catkins and old cones (February)

Twig in February

Tree in February

Leaves and young (next year's) catkins in September

Grey Alder *A. incana* is similar but its leaves are a more 'normal' oval shape (**widest around the middle** and with more **pointed tips**) and have **greyish undersides**, and the **twigs have short hairs** (hairless in Alder). Leaf similar to that of Whitebeam but underside greyish (not whitish) & twigs thinner. **Planted** (and locally naturalised) in **urban and rural situations**, mainly in the lowlands.

Italian Alder *A. cordata* has **glossy leaves** with **heart-shaped bases**, and **larger 'cones'** (up to 3 cm long). **Planted** in **urban and rural situations**, mainly in lowlands (as in Grey Alder, but not as common than that species).

Three common Willows (*Salix* species)

Willows have **slender twigs, oval leaves** and, in spring, **pale fluffy catkins**. Commonest on **moist to wet ground**. Shoots **palatable** to animals such as sheep, deer and cattle. Three very common species described on this page; more overleaf.

◄ **Grey Willow** *Salix cinerea*: **multi-stemmed, densely-branched shrub** to 5 m tall; **narrow, pointed, toothed leaves** to 8 cm long, 2-3 times as long as wide and **widest beyond the middle**. Autumn leaves yellowish or mottled green, yellow, brown & black. Small oval buds held close against the twig. On **damp or wet, acid to neutral soils**; in woods mainly in W1-7 & W18.

Eared Willow ► *S. aurita*: similar to Grey Willow but shorter (to 3 m); twigs stick out from larger twigs/branches at wide angles; **leaves shorter** (to 3 cm) and 1.5-3.0 times as long as wide, **greyer green** with **broad blunt tips and wrinkled texture**. On **damp or wet, acid to neutral soils**, especially in upland areas; in woods mainly W4, W7 & W18.

Goat Willow stipules present or, esp. in N, absent

◄ **Goat Willow** *S. caprea*: **tree** (or if a shrub, less densely multi-stemmed than last two species) with **broad oval, toothed leaves** up to 12 cm long, 1-2 times as long as wide, and **widest at middle**. Small oval buds stick out from twig. On **dryish to damp, neutral to acid soils**, especially in woods (W7-11).

Five more Willows (*Salix* species)

Bay Willow *Salix pentandra* is a **shrub or small tree** with **oval leaves** to 12 cm long, 2–3 times as long as wide, widest around the middle and with a **glossy upperside**. Mainly on **wet neutral soils** from English Midlands northwards; in woods mainly in W3.

Crack Willow *S. fragilis s.l.* * is a **tree** that can be **large, with widely-spreading branches. Twigs are brittle at their junctions**. Leaves are **narrow oval, finely-toothed**, to 15 cm long, 4–9 times as long as wide. Mainly on **wet neutral soils in lowlands**, especially **by water**; in woods mainly in W6. **Not native.**

White Willow *S. alba* is similar to Crack Willow but **twigs not brittle**, leaves **shorter** (to 12 cm), 5–8 times as long as wide and with **whitish undersides**. Mainly on **wet neutral soils in lowlands**, especially **by water**. **Not native.**

Osier *S. viminalis*: a **shrub** with **long thin stems**, **very long, very narrow, untoothed leaves** to 25 cm long, 7–18 times as long as wide. On **wet neutral soils in lowlands**, especially **by water**. **Not native. Stems – especially from coppiced Osiers – can be used for basket-making.**

Purple Willow *S. purpurea* is a shrub with **narrow oval leaves** to 8 cm long and 4–15 times as long as wide; some leaves in **nearly opposite** pairs. On **wet ground**, mainly in lowlands. Native, but also **commonly planted**.

Crack Willow (July)

White Willow (July)

Bay Willow (June)

Osier leaves (September)

* = *sensu lato* = in the broad sense: probably *S. x fragilis*, a *S. alba* x *S. euxina* hybrid.

Summary table of identification features of the commoner Willows

The species are ordered here by the average leaf length : leaf width ratio (second column from left), which shows how many times longer than wider the leaves are.

	Leaf length divided by leaf width	Maximum length of leaf	Additional notes
Goat Willow *Salix caprea*	1-2	12 cm	Bush or tree; leaf broad oval, widest at middle
Eared Willow *Salix aurita*	1.5-3.0	3 cm	Dense bush; many twigs stick out at a wide angle; leaf oval and widest beyond middle, with broad rounded tip and wrinkled texture
Grey Willow *Salix cinerea*	2-3	8 cm	Dense bush; leaf narrow oval and widest beyond middle; pointed tip
Bay Willow *Salix pentandra*	2-3	12 cm	Bush or small tree; leaf glossy above
White Willow *Salix alba*	5-8	12 cm	Tree; leaf whitish beneath
Crack Willow *Salix fragilis*	4-9	15 cm	Big tree; twigs brittle at junctions
Purple Willow *Salix purpurea*	4-15	8 cm	Bush; some leaves nearly opposite
Osier *Salix viminalis*	7-18	25 cm	Bush with very long straight branches; leaves untoothed

Lower stems of Grey Willow in wet woodland (NVC W5) in East Lothian, Scotland

Poplars *Populus* species

Fast-growing trees with **rather thick twigs with oval, pointed buds**, **leaves on long stalks**, and thick, hanging **catkins** in spring. Leaves mostly turn yellow in autumn. Commonly reproduce by **suckers** arising from spreading underground roots.

Aspen *P. tremula*. **Round, wavy-edged, mid green leaves** on long stalks that are flattened (from side to side) and flimsy, so leaves **rattle** easily in wind. Native on **dry to moist neutral to acid soils**, mainly in **woods** (esp. W11) and on **cliffs**. Young growth **palatable** to sheep, deer etc. Leaves on young lower shoots can be pointed.

White Poplar *P. alba* has **leaves with dark green uppersides** and **white undersides; triangular with irregular blunt teeth, or 3-5 lobes in a star-like pattern. Young twigs are white. Introduced & naturalised**, mainly on neutral soils in the lowlands.

Grey Poplar *P.* x *canescens*; an **Aspen/White Poplar hybrid; oval-triangular leaves with irregular, blunt teeth, dull green uppersides & greyish undersides. Planted** in lowlands.

Hybrid Black-poplar *P.* x *canadensis* has **triangular, toothed, pointed, mid green leaves** (like big birch leaves). **Leaf stalk flattened**, with **1-2 small warts** where it joins leaf blade. **Black Poplar** *P. nigra*: similar but **leaf stalk lacks warts**, **branches arch downwards** and **old trunk is 'knobbly'**. **Lombardy Poplar** *P. nigra 'italica'*: like Black P. but **unmistakably tall & thin**, with smaller leaves. **Balsam-poplars** *P. balsamifera, P. trichocarpa* + hybrid: like Hybrid Black-p. but leaf stalk **not flattened** and variably hairy; buds **smell of balsam**. All **introduced & planted** in **lowlands**, except Black P., which is a rather scarce native.

Aspens in winter

Hybrid Black-poplar (May)

Lombardy P.

Aspen leaves

Grey Poplar leaf

White Poplar leaves

Wild cherry *Prunus avium*

Also called **Gean** (pronounced *'Jee-un'*). A small to medium-sized tree whose smooth brown bark has **horizontal lines** (bark peeling off). The thick grey-brown twigs (>6 mm thick) have oval, pointed, brown buds. The large **oval, toothed leaves** (up to 16 cm long) are on **medium-length stalks** (up to 5 cm long), and there are **two small reddish glands** (looking like warts) where the base of the leaf blade joins the top of the leaf stalk. Leaves turn red or orange in autumn. Clusters of long-stalked **white flowers** in spring become **edible cherries** in summer. On **dry to moist, neutral soils**; in woods mainly W8-11.

Wild Cherry leaves in May

Flowers in April

Trunk

Bird Cherry *P. padus*: a shrub or small tree with **thinner, darker twigs, smaller leaves** (<10 cm long) on **shorter** (<2 cm) **leaf stalks** (with smaller glands); **flowers/fruits in long, narrow clusters** (racemes). On **damp, neutral to basic soils**, especially in **woods** (W7/9) in **N Britain**.

Bird Cherry in May

Dwarf Cherry *P. cerasus*: like Wild Cherry but **leaves smaller** (<9 cm) & **darker**; **small groups of fewer flowers** on **shorter stalks**; **wider-spreading petals**; **twigs thinner & dull grey-brown**; **planted** in lowlands.

Cherry Plum *P. cerasifera*: rather like Dwarf Cherry but **leaves smaller** (<7 cm long), **flowers earlier** (Mar-Apr), **twigs shinier and green** (or with some red). **Planted** in lowlands. Compared with Blackthorn (page 39) leaves and flowers are larger and twigs are not thorny.

Apples, Pears and Plum/Damson

Small fruit trees with **oval, stalked leaves with toothed edges**, and **white flowers** in spring developing into the familiar **edible fruits** in late summer and autumn.

Crab Apple *Malus sylvestris* has **thick twigs** that can have some **thorns**, its **leaves are mid- or yellow-green, hairless** and up to 5 cm long, and its apples are yellow-green and up to 2 cm across. **Native** in **woods, scrub and hedges**.

Apple *Malus pumila* is similar but **not thorny**, has leaves up to 15 cm long and with **hairy undersides**, and variably coloured apples up to 12 cm across. **Introduced**. In **gardens, orchards and hedges**.

Wild Pear *Pyrus pyraster* has **rough bark with squared pattern**, thick, **thorny** twigs, **mid- or yellow-green leaves** up to 7 cm long with **glossy, hairless uppersides**, and **sour-tasting** pears up to 4 cm across. **Introduced**. In **woods & hedges**.

Pear *Pyrus communis* is like Wild Pear but **without thorns** and with **sweet-tasting** pears up to 4 cm across. **Introduced**. In **gardens, orchards and hedges**.

Plum/Damson *Prunus domestica* has **dark bark, thin twigs, mid to dark green leaves** up to 8 cm long, and **fruits very varied** and including **plums** (ssp. *domestica*) to smaller purple **damsons** (ssp. *insititia*). Damson twigs can have some thorns; plum twigs lack thorns. **Introduced**. In **gardens, orchards and hedges**.

Apple

Pear

Plum

Damson

Holly *Ilex aquifolium*

A small tree with **smooth grey bark, green twigs** and **dark, stiff, evergreen leaves with very sharp spines at the ends of side lobes (some spines pointing up; others pointing down) and at the tip**. Leaves in ungrazed places (e.g. higher up trees) can lack side lobes but still have a spiny tip. Leaf upperside very **glossy**. **Small white flowers** in spring. **Red berries** in autumn-winter. Grows on **dry to moist, acid to neutral soils**, especially in **woods** (W10–11 & 14) and **hedges**. Regeneration mainly within woodland. **Palatable** to animals such as sheep and deer; young holly in grazed woods commonly bitten back to low 'bushes' <1 m tall.

Leafy shoot

Shoot with berries

Young holly in oak woodland that has been ungrazed for the last 20 years

Highclere Holly *I. x altaclarensis* (hybrid with the Macaronesian species *I. perado*): **leaves relatively wider**, with **spines relatively shorter and mostly in one plane, pointing straight out or slightly downwards**. Mainly **planted** in various **lowland rural and urban habitats**.

Evergreen Oak *Quercus ilex* has **similarly evergreen, rather leathery leaves** that are oval (with or without side lobes), but compared with Holly its **leaves are less stiff** and **less glossy**, and any spines along their edges are **shorter and not stiff or prickly**. It is a **larger tree**, with **rougher fissured bark** and is commonly **planted**, esp. in **parks/gardens**.

Sweet Chestnut *Castanea sativa*

A tree with **fairly thick twigs that are angular in section** with ridges running along them and with many tiny creamy-white dots. Small green-brown buds, **long, oval, sharply toothed leaves**, and (in autumn) round, green, **very spiny fruit cases** enclosing clusters of chestnuts (**edible if they can grow large enough**). Leaves turn brown or orange-brown in autumn. Old thick trunks can be gnarled and have deep furrows running up at an angle. **Introduced.** Mainly in lowlands; **coppiced** in many woods (W10) in SE England.

Leaves (August)

Young fruit & old flower catkin (August)

Blackthorn *Prunus spinosa*

Dense shrub with **dark, spiny twigs**, small buds (commonly clustered in winter) and **small, narrow, oval leaves up to 4 cm long. White flowers** in early spring **before** leaves; **dark blue, edible but sour-tasting berries (sloes)** in autumn. On **dry to moist neutral to basic soils** at low altitudes, in **hedges** and **woods** (W8-11) and forming dense patches of **scrub** (W22).

Flowers in April (before the leaves)

Leaves and berries in September

Hawthorn *Crataegus monogyna*

A **shrub or small tree** with **brown or red-brown, spiny twigs, small buds** (purplish-red-tipped in winter) and **small oblong or triangular, deeply lobed leaves** whose edges are cut at least halfway to the central vein of the leaf. Leaves **edible** when fresh in spring. Clusters of **whitish flowers** in spring; **red berries** in autumn. On **dry to moist, acid to basic soils**, forming dense **scrub** (W21) and **hedges**, or scattered in **woods** (esp. W8–10) and **grasslands**.

Leaves and flowers in May

Leaves and berries in September

Hawthorn hedge

Hawthorn trees

Scattered Hawthorns on hillside

Midland Hawthorn *C. laevigata* is similar but the **leaf margins are cut less than halfway to the central vein** and the **side veins curve toward the leaf tip**. In **woods** (esp. **ancient** W8/10), **scrub** (W21) and **hedges** on **neutral soils** in **central and SE England**.

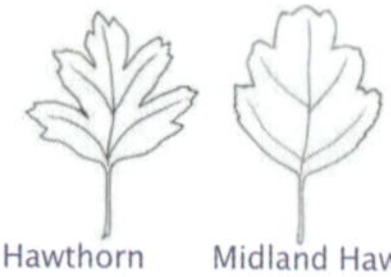

Hawthorn Midland Hawthorn

Sea-bucktorn *Hippophae rhamnoides*

A shrub up to 4 m tall, with a dense mass of spiny branches and twigs, greyish-green, very narrow leaves with untoothed edges, and clusters of orange berries that remain on the shoots through the winter. Grows on sandy ground by the sea. Native along the E coast from Kent to Northumberland (except N Yorkshire & Co. Durham). Introduced elsewhere. Spreads by means of underground shoots (rhizomes) from which new bushes arise: in this way it can form dense, invasive thickets.

Leaves and berries in September

Bushes in winter

Gooseberry *Ribes uva-crispa*

A shrub up to 1.5 m tall, whose green (young) or brown (older) twigs have sharp spines, stalked leaves with 3 main lobes (each lobe with toothed edges and a bluntish tip) and small pale greenish or reddish flowers (spring) becoming green edible gooseberries during summer. Introduced. Mainly on dry to moist, neutral soils within and at the edges of lowland woods (esp. W8 & W10), and in hedges.

Leaves and young berries in May

Rhododendron *Rhododendron ponticum*

A **shrub** up to 4 m tall, with **thick twigs** and **large, dark green, leathery, short-stalked, oval, untoothed evergreen leaves spreading out in all directions**. Large **pink-purple flowers** in early summer. Cherry Laurel looks similar but has **alternate, toothed leaves** and **small white flowers**. **Introduced and naturalised** on **acid soils**, especially in western **woods**; also in **heaths** and **bogs**. An aggressive colonist forming dense thickets whose overshading and leaf litter cause severe losses of ecological interest. **Eradication needed in many western districts**. (Most of the other Rhododendron species planted in gardens do not spread aggressively.)

The pink flowers are pretty, but from a conservation viewpoint **removal of Rhododendron** is preferable, as shown in the above example removed using the Lever & Mulch method developed in Morvern in W Scotland - see page 350).

Rhododendron: untoothed leaves sticking out from the stem in all directions

Cherry Laurel: toothed leaves in two rows (left & right) along the stems

Broom *Cytisus scoparius*

A shrub 1-3 m tall, with **green, ridged stems** (as in Bilberry - page 62) and **yellow, pea-like flowers** (as in Gorse - page 44). **Not spiny**. Lower leaves have **3 oval leaflets up to 2 cm long**; uppermost leaves simple and narrow oval. On **dry, acid to neutral soils**, in **grasslands, heaths, woodland margins** and on **riverbanks** and **disturbed ground** (e.g. banks at **edges of roads and tracks**).

Shoot in April

Flowering bush in June

Seed pods in September

Leaf in October

Dyer's Greenweed *Genista tinctoria* is like a small Broom <70 cm tall. **Leaves simple, narrow, oval, to 3 cm long**. Yellow flowers in summer. Seed pods very dark. In **calcareous to slightly acid grassland, heath and wetland** in Britain N to southernmost Scotland.

Petty Whin *G. anglica* is similar but **stems have spines up to 2 cm long**, and **smaller oval leaves <1 cm long**. Yellow flowers (in spring) develop into swollen brown seed pods. Old stems are brown. In **acid heaths and grasslands** in England, Wales and E Scotland.

Dyer's Greenweed

Petty Whin

Gorse *Ulex europaeus*

Also called **Whin** (especially in Scotland) and **Furze** (especially in southern England). A shrub **up to 3 m tall**. Stems clothed with dense, dark masses of **very sharp, deeply grooved, evergreen spines** of a greyish-green colour. Conspicuous **yellow, pea-like, coconut-scented flowers**, mainly in March–June but can occur at any time of year. **Calyx** (see flower photo) **is 10–17 mm long**. **Bracteole** (small flap at base of flower or, later, seed head – see photo below) is **3–5 mm wide**. On **dry, acid to neutral soils**, in **grasslands, heaths, woodland margins**, and on **riverbanks** and **disturbed ground**. Forms dense patches (W23). Commonly **browsed** by animals such as sheep and deer. Can threaten conservation interest in habitats such as herb-rich grassland: **controllable by cutting**, but **burning can help it spread** in the long term.

Gorse in April

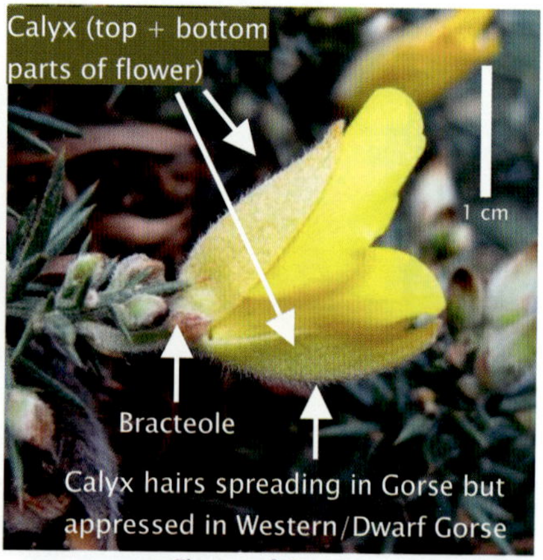

Flower of Gorse

Western Gorse and Dwarf Gorse

These look like Gorse but are **shorter**, and their **spines are dark green** (not greyish-green) and **faintly grooved** (not deeply grooved). They **flower in July–September** and the **bracteole** (see flower photo above) **is only 0.6 mm wide**.

Western Gorse *U. gallii* (see photo on page 358) is up to **2 m tall** and the **calyx** (see photo above) **is >9 mm long**; on **dry to damp, mainly acid soils** in **heavily to lightly grazed heaths** (H4–6/8 & M15) and **grasslands** (U4/5) N to S Scotland. Commonest in SW England, Wales & Ireland. Scarce non-native in N Scotland.

Dwarf Gorse (also called **Dwarf Furze**) *U. minor* is **<1 m tall** and the **calyx** (see flower photo) **is less than 9 mm long**; on **dry to damp, acid soils** in **lowland heaths** (H2–3) **in S–SE England** and, less commonly, in Nottinghamshire and N Cumbria.

Section 1b

CONIFERS

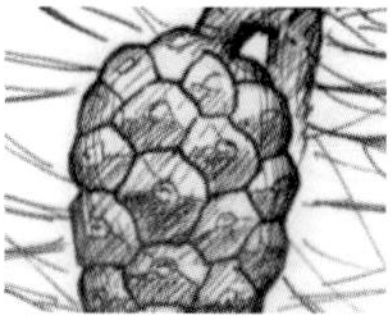

These trees and shrubs have needle-like or scale-like leaves that are evergreen except in the case of the Larches. Their 'flowers' are not true flowers in the strict sense; in most species they develop into woody cones. Most conifer species found in Britain & Ireland are introduced here; only Scots Pine, Yew and Juniper are native. For identification, first assign a plant to one of these five groups:

Needle-like leaves growing **SINGLY** along the twigs (not in pairs or groups)		Spruces Firs Yew Western Hemlock
Needle-like leaves in **PAIRS,** joined at their bases by sheaths	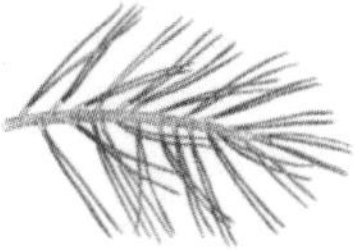	Pines
Needle-like leaves in **THREES**, spreading out in different directions		Juniper
Needle-like leaves in **CLUSTERS** (except at shoot tips, where they're single)	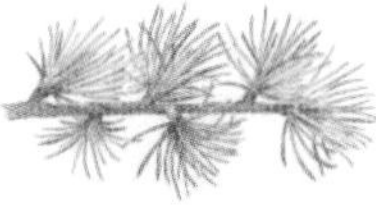	Larches Cedars
SCALE-LIKE leaves **OVERLAPPING** each other		Cypresses Wellingtonia Western Red Cedar

Page 46 deals with the identification of 21 conifer species but does not have room for their Latin names, so these are given below, with their English names (species in the same order as on page 46).

Norway Spruce	*Picea abies*	Juniper	*Juniperus communis*
Sitka Spruce	*Picea sitchensis*	European Larch	*Larix decidua*
Western Hemlock	*Tsuga heterophylla*	Japanese Larch	*Larix kaempferi*
Douglas Fir	*Pseudotsuga menziesii*	Cedar of Lebanon	*Cedrus libani*
Yew	*Taxus baccata*	Deodar	*Cedrus deodara*
Silver Fir	*Abies alba*	Mount Atlas Cedar	*Cedrus atlantica*
Grand Fir	*Abies grandis*	Lawson's Cypress	*Chamaecyparis lawsoniana*
Noble Fir	*Abies procera*	Leyland Cypress	*x Cupressocyparis leylandii*
Scots Pine	*Pinus sylvestris*	Western Red Cedar	*Thuja plicata*
Lodgepole Pine	*Pinus contorta*	Wellingtonia	*Sequoiadendron giganteum*
Corsican Pine	*Pinus nigra*		

Pages 47-49 give descriptions and photographs of our three native conifers.

Identification of the commoner conifers

Evergreen needle-like leaves growing SINGLY along twigs (not in pairs or groups); cone direction varied

Species	Leaf length (mm)	Underside of leaf compared with dark upperside	Leaf texture	Leaf tip	cone length (cm) and direction
Norway Spruce (P)	10–20	similar dark green	stiff	pointed	12–15 (down)
Sitka Spruce (P)	20–30	much paler *	stiff	pointed	5–8 (down)
Western Hemlock	6–20	much paler *	soft	pointed	2–3 (down)
Douglas Fir	20–25	much paler *	soft	pointed	5–8 (down)
Yew (N)	20–40	paler (pale green)	soft	pointed	(red 'berry')
Silver Fir	12–20	much paler *	stiff	blunt	10–15 (up)
Grand Fir	20–50	much paler *	stiff	blunt	5–10 (up)
Noble Fir	15–35	similar blue-green	stiff	blunt	14–25 (up)

Extra notes: (1) spruce leaf bases grow from projecting woody 'pegs' on the twigs; (2) cones of the last three species disintegrate on the tree, their scales falling to leave an upright bare central column; (3) Western Hemlock leaves can be very variable in size within the same shoot.

Evergreen needle-like leaves in PAIRS joined at their bases by sheaths; cones point in all directions

Species	Leaf length	Leaf colour	Trunk
Scots Pine (N) (P)	3–8 cm	blue-green	grey + some red-brown
Lodgepole Pine (P)	3–8 cm	dark green	dark greyish (no red)
Corsican Pine (P)	8–16 cm	dark or grey-green	greyish (no red)

Evergreen needle-like leaves in THREES Juniper (N) – a shrub with stiff, pointed (prickly), unsheathed leaves with dark uppersides and pale undersides; leaves of each group of three stick out in three directions from same point along the twig; cones like berries, green at first, then black.

Needle-like leaves in dense CLUSTERS (except at shoot tips, where they're single); all cones point up

Species	Notes
European Larch (P)	not evergreen; leaves light green; twigs yellowish; cone-scales straight
Japanese Larch (P)	not evergreen; leaves blue-green; twigs reddish; tips of cone-scales curl out
Cedar of Lebanon	evergreen (leaves dark green); several trunks; branches horizontal
Deodar	evergreen (leaves dark green); one trunk; branches horizontal + drooping tips
Mount Atlas Cedar	evergreen (leaves blue-green); one trunk; branches slant upwards

Evergreen TRIANGULAR, SCALE-LIKE leaves OVERLAPPING each other; cones small – directions varied

Species	Shoot shape	Shoot underside colour **	Leaves smell of:	Cone shape	Cone direction
Western Red Cedar	flattened	streaked whitish	pineapple	oval	sideways
Lawson's Cypress	flattened	pale green	parsley	round	sideways
Leyland Cypress	not flattened	yellow-green	parsley	round	sideways
Wellingtonia ***	not flattened	mid-dark green	leather	oval	down

* Whitish-green (green + two white or whitish lines)
** All uppersides mid to dark green
*** Unmistakeable big tree with thick trunk and thick, soft, spongy, reddish-brown bark
(N) Native, at least in part of its British range; all other species on this page are introduced
(P) The main species of commercial conifer plantations in Britain & Ireland

Scots Pine *Pinus sylvestris*

Identified by the **blue-green** needle-like leaves **3-8 cm long, sheathed in pairs**, and the bark a mixture of **grey** and **red-brown**. Native on **dry to wet acid soils** through much of the **Highlands** (dominant in W18 woodland); **planted elsewhere** and within native range. Regeneration mainly in open or lightly shaded situations; can be dense and quite even-aged; **often browsed** by deer and sheep; grows best where grazing is no more than light. Lodgepole Pine (planted on damp soils in uplands) is similar but with dark green (not blue-green) leaves, and a grey-brown trunk with no red colour. Corsican Pine (planted commonly, especially on dry sandy soil in lowlands) has longer (8-16 mm) leaves with no blue tinge, and no red colour in the bark.

Scots Pine in the eastern Scottish Highlands

Shoots with leaves and unopened cone

Fallen shoot and opened cone

PL = pair of leaves; **Sh** = sheathed leaf bases

Yew *Taxus baccata*

A **dense tree** with narrow, **flattened** leaves that are 2-4 cm long and commonly curved down at the tips. The leaves are borne singly along the twigs (not in pairs or groups). Each leaf has a dark shiny upperside and a **more or less uniformly pale green underside**; it lacks the strong pattern of a central dark stripe and two whitish side stripes as seen on the undersides of the leaves of Sitka Spruce, Western Hemlock, Douglas, Silver Fir and Grand Fir. The **cones look like red berries**. Yew is **native in woods** (W12-13) **and on limestone cliffs** in parts of England, Wales & Ireland. It is also **commonly planted** throughout Britain & Ireland, especially in the lowlands. All parts of the plant except the red fleshy fruit-case are **highly poisonous**.

Upperside of shoot

Underside of shoot

Yew growing wild in the Lake District

Shoot with red 'berries' (Sept)

Juniper *Juniperus communis*

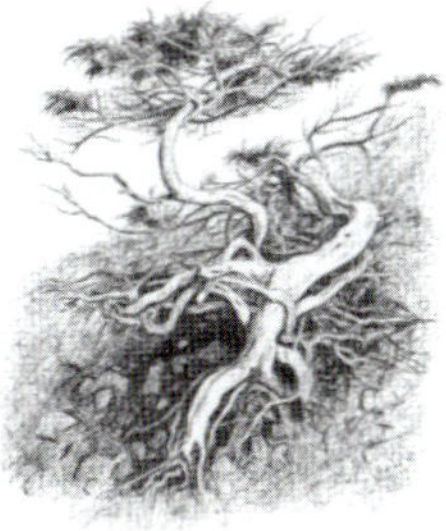

Old Juniper bush in heathland

Juniper has a distinctive **shrubby** growth form, and its needle-like **leaves are in groups of three**. The cones look like small **berries** (**green first, then dark blue** - **used to make gin**). Subspecies *communis* grows **to about 4 m** (columnar or spreading), with **widely spreading, sharply pointed leaves**. It forms patches of **scrub** (W19) on **dry to moist acid to basic soils** on **heathy or grassy hillslopes** in central & E Highlands, Lake District, Southern Uplands and N Pennines; it also occurs in mixed scrub (W21) on chalk in SE England. Subspecies *nana* (Dwarf Juniper) is **low** (mostly <20 cm) and **prostrate**, with **blunter, less spreading leaves**; on **thin acid soils** in **exposed, coastal to mountainside rocky heaths** in W Highlands, Hebrides & Northern Isles; also in W Ireland and on some hills in the Lake District and NW Wales. Juniper is readily **grazed** by sheep and deer and is also **vulnerable to burning**. It is a good **indicator of unburned or little-burned habitats**.

Ssp. *communis*

Ssp. *nana* (with 'berries')

Juniper scrub (ssp. *communis*)

Heath of Dwarf juniper (ssp. *nana*) & Heather

Section 1c

DWARF SHRUBS

These are shrubs with woody, branched stems growing up to about a metre in height. Some species are much shorter, as are potentially tall species growing in exposed or heavily grazed places or on very thin or nutrient-poor soils.

The group is made up mainly of Heathers and the related Bilberry, Cowberry, Crowberry and Bearberry: species found most commonly in heaths and bogs.

Heather *Calluna vulgaris* is particularly common and exhibits a wide range of growth forms in response to grazing, burning and wind-exposure. This makes it an important species to look at when assessing the condition of vegetation and making decisions about land management, so seven additional pages are provided to describe and illustrate variation in its condition.

Five woody species that grow as low dwarf shrubs in heaths and grasslands are described elsewhere in this book: the prostrate form of Juniper (see page 49 in the Conifers section), Dyer's Greenweed, Petty Whin, Western Gorse and Dwarf Gorse (described with Broom and Gorse on pages 43 and 44 in the Broadleaved Trees & Shrubs section).

Tutsan is described in this section but is also included with six non-woody St John's-worts on page 244.

Thyme and Rockrose are dwarf shrubs that can look like herbs because of their bright flowers and low growth form. However, their stems are woody so they are dwarf shrubs and are therefore described in this section.

Sea-purslane (see page 294) is also shrubby, with branched stems up to 80 cm tall (woody in their lower parts) but distinctive with its fleshy, evergreen, grey-green, narrow oval leaves up to 5 cm long, and its saltmarsh habitat.

Dwarf Willow is a very low-grown dwarf shrub: see page 370.

Heather *Calluna vulgaris*

Also called **Ling**. A woody dwarf shrub with **tiny evergreen 3-sided hairless or hairy leaves 1-2 mm long**, in **opposite pairs** along the shoots, forming **4 vertical rows of leaves along younger shoots** where pairs of leaves (alternately at 90° to each other) are densely overlapping (see photo on right). The **pale pink-purple flowers are small (4 mm long) and in spikes** at the shoot tips in mid to late summer. Leaves & flowers are shorter than in Bell Heather and Cross-leaved Heath (pages 59-60).

Heather grows on **dry to wet, mainly acidic soils, including peat**. It is abundant to dominant in **heaths** (H1-17, H21-22 & M14-16) and **bogs** (M17-19 & 21). It also grows in **woods** (W16-18) and **juniper scrub** (W19), and sparsely in **grasslands** (some examples of U2-6 and CG10-11).

Heather is **palatable** to animals such as sheep, deer and cattle (and also rabbits), and is therefore a **useful indicator of grazing intensity**. Ungrazed plants can be up to 1 m tall with fairly straight shoots and abundant flowers. Grazed plants are shorter (can be <5 cm), with bitten-off shoot tips easily visible, and with fewer flowers. Various growth forms that develop in response to grazing and burning are shown in the following pages. Bees can make **heather honey** from heather nectar.

Heather flowers in July

Heather condition - page 1/7

Lightly grazed heather: tall (60 cm) and straight-grown, with abundant flowering. Flowering now over but old flowers still present and easily visible in this photo taken in the Lammermuir Hills, East Lothian in December.

Moderately grazed heather 30-40 cm tall, quite straight-grown and flowering well, but rather varied in height and with a slightly 'clumped' appearance: Dundreggan, Inverness-shire, in August.

Heather condition - page 2/7

Heather that was once only lightly grazed and therefore quite tall, but has in recent years been more **heavily browsed**, so the short, dense outer shoots give the plant a **topiaried** look: this indicates **recently increased grazing pressure**. Lammermuir Hills, East Lothian, in December.

With more **prolonged heavy grazing** a further development from topiaried Heather is this **drumstick** growth form that also indicates **increased grazing pressure in recent years**. Lammermuir Hills, East Lothian, in May. Note how the heavy grazing has also produced grass-dominated vegetation (acid grassland) here.

Heather condition - page 3/7

This is old Heather that has grown tall and straight (only lightly grazed) until **recently increased sheep densities have resulted in trampled, broken stems and very short, heavily grazed associated grass swards**. Very heavy grazing can cause heather to be replaced by grasses. Lammermuir Hills, East Lothian, in May.

Heather **very heavily browsed** into a small, very low, dense clump: this indicates **very high grazing pressure, probably for several years**. No sign of flowering. Associated vegetation has very little Heather (mostly Bilberry and grasses - Heather has been grazed out). Lammermuir Hills, East Lothian, in May.

Heather condition - page 4/7

Heather grazed into a **low carpet**: this indicates **long term heavy grazing pressure**. Near Blair Atholl, Perthshire, in June.

Close view into the bottom of a fairly tall Heather bush whose lower stems (seen here) are contorted as a result of **heavy grazing in the past**, even though more recent growth (not visible here - above top of photo) is straighter and not heavily grazed. This is an indication that **grazing is now lighter than it used to be**.

Heather condition - page 5/7

Fairly short heath in which Heather and Bell Heather are dominant but grasses are common and mixed evenly through the sward. This is the result of **moderate grazing throughout the year, spread evenly over a large area**. Perthshire, in July.

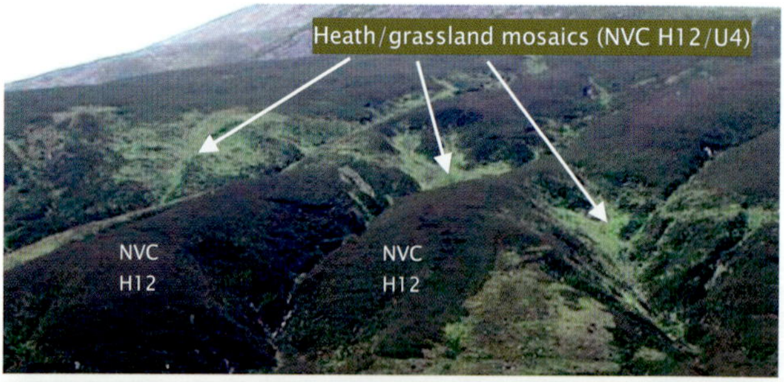

Heavy grazing concentrated in sheltered places, especially in winter, produces these patchwork mosaics of short grassland (NVC type U4 in these photos) and contrasting taller, darker, dense clumps of Heather and dense patches of heath (H12). Perthshire, in July.

Heather condition - page 6/7

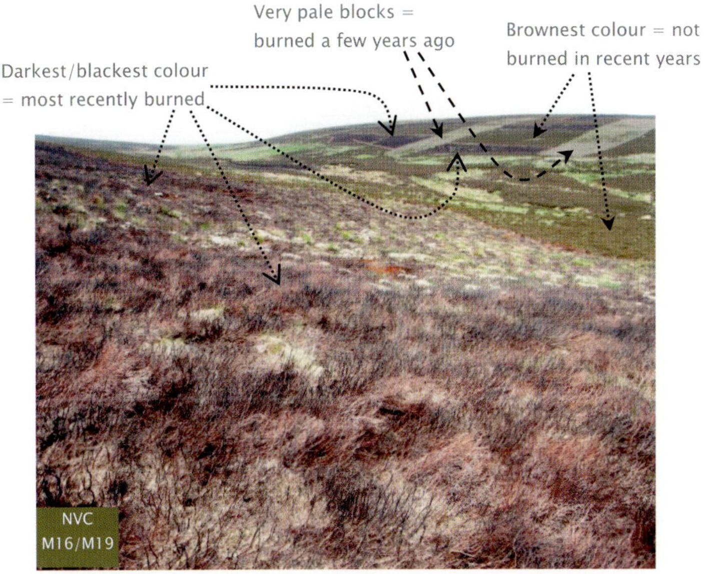

Recently burned Heather, with patchwork effect from burns in previous years visible in the distance. See page 359 for discussion about heather burning. Lammermuir Hills, East Lothian, in May.

Young, short Heather **growing back since heath was burned** 2-3 years ago. Very pale old burned Heather stems still present. Lammermuir Hills, East Lothian, May.

Heather condition - page 7/7

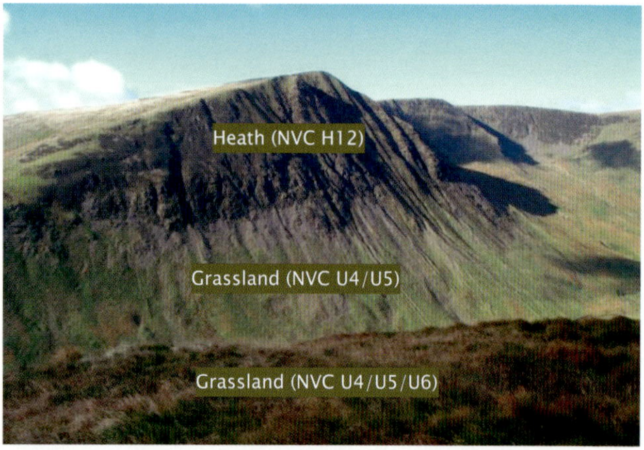

Heather abundant **only on steep, less accessible slopes** that are **lightly grazed or ungrazed** (NVC H12): the more **accessible slopes are heavily grazed** and support grassland (U4/5/6) with little or no Heather. Blackhope, Southern Uplands, October.

Heather that is **very short and prostrate** because of **frequent exposure to strong wind**. This growth form is found mainly on exposed ground at medium to high altitude in the Scottish Highlands, and at lower altitudes in the extreme NW Highlands, the Hebrides, Orkney and Shetland. Beinn a' Ghlo, Perthshire, in July.

Bell Heather *Erica cinerea*

Rather like Heather (page 51) but with **longer hairless leaves (5-7 mm) in groups (whorls) of 3**, and with **fewer, larger (5-6 mm), bright pink-purple flowers**. Where the leaves grow from the stem, there can also be very short branches with smaller leaves. Bell Heather resembles Cross-leaved Heath (page 60), but its leaves are hairless and in groups of 3 and the flowers are darker and in more elongated spikes. Old flowers turn orange-brown in autumn and can remain right through the winter. Bell Heather grows mainly on **dry to moist acid soils**, especially in **dry heaths** (H2-8, H10 & H14-16), but also in some **wet heaths** (M15c) and in **woods** of birch/oak (W17) and pine (W18). It is most common in the west, at low to moderate altitudes. It is **palatable** to grazing animals in the same way as described for Heather.

Bell Heather in flower in July

Dorset Heath *E. ciliaris* is rather similar but has **short (<3 mm) hairy leaves**; **stems also hairy**. Rare in **dry and wet heaths** in S England (Cornwall E to Dorset).

St Dabeoc's Heath *Daboecia cantabrica* has **larger flowers up to 12 mm long**, and **oval leaves** up to 1 cm long and with **whitish undersides**. Locally common in **dry heaths** in mid-W Ireland.

Cross-leaved Heath *Erica tetralix*

This resembles Bell Heather (page 59) in having narrow leaves in groups (whorls), but it is easily separated by the leaves being **grey-green** in colour (they have white **hairs** on them, giving the whole plant a grey tinge), **shorter** (2-4 mm) and in **groups of 4** (not 3), and also by the **paler pink flowers** in **short round clusters at shoot tips**. As in Bell Heather the old flowers become brownish in colour and can persist through the winter. Cross-leaved Heath prefers wetter ground, growing mainly on **moist to wet, acid, peaty soils** in **wet heaths** (M14-16 & H3-7) and **bogs** (M17-19 & 21), and in some wet woods (W4/18). It is **not very palatable** to grazing animals such as sheep and deer; it commonly survives in quantity in heaths and bogs where Heather has been eliminated by heavy grazing; grazed plants of Cross-leaved Heath indicate that the general grazing intensity is high.

Cross-leaved heath in July, with some old, brown flowers in the background

Crowberry *Empetrum nigrum*

The **narrow, hairless, shiny dark green, evergreen leaves** 4-6 mm long look rather like those of Bell Heather but are **thicker**, arranged spirally along the shoots (or with some whorls of 2-3) and have a **white stripe on the underside**. The flowers (May-June) are tiny and inconspicuous, becoming **round black berries** in late summer. Forms **low patches and trailing shoots** on **dry to moist, acid soils** in **dwarf shrub heaths** (H7 & H10-22), **montane moss heaths and snowbeds** (U7 & U10) and **bogs** (M18-19): mostly in **short vegetation** in **cool/cold northern and western places**; down to sea level in the N. **Unpalatable**: grazed plants indicate high grazing intensity. There are two subspecies (can be difficult to identify to subspecies level):

- **ssp. *nigrum*** (the **commonest** subspecies) has oblong, parallel-sided leaves **3-4** times as long as wide, and the young stems are **reddish**
- **ssp. *hermaphroditum*** (mainly on **mountains** in the Scottish Highlands) has oblong-oval leaves **2-3** x as long as wide, and the young stems are **green**

Shoots viewed from above

Undersides of leaves, showing white stripes

< Shoot with berries (July)

Trailing Azalea *Kalmia procumbens* ► leaves similar but **no white underside stripe**; **central vein runs in a sunken groove on upperside**. Plant **very low and creeping** (small patches), 5-petalled **pink flowers** (May-July). In **exposed montane heaths** (H13-20, U9/10) in the Highlands, Skye, Orkney & Shetland.

Bilberry *Vaccinium myrtillus*

Also called **Blaeberry** in Scotland and **Whortleberry** in SW England. A woody dwarf shrub with **unmistakeable green, ridged stems** bearing **oval, toothed, pointed, deciduous leaves 1–3 cm long.** It has pale pink-green flowers in spring, becoming **edible dark blue berries** in autumn. Leaves turn reddish in autumn. Mainly on **dry to moist, acidic soils**, in **dry heaths** (H8 & 12–22) and **woods** (W16–18), but also in some **wet heaths** (M15–16), **bogs** (M17–20), **grasslands** (U2 & 4–7, & CG11), **mountain moss heaths** (U10), **cliff ledges** (U16) and among **bracken** (U20) and **juniper scrub** (W19). More **tolerant of cold** (including snow cover) **and shade** than Heather. **Palatable** to grazing animals such as sheep and deer: a useful indicator of **grazing** intensity. **Ungrazed** plants can be up to 1 m tall with fairly straight shoots and abundant flowers/fruits. **Grazed** plants are shorter (can be <5 cm), with many bitten-off shoot tips and with fewer flowers/fruits. Can be more tolerant than heather of grazing and burning: its creeping underground shoots (rhizomes) may help in this respect.

Grazed Bilberry heath (NVC H18) (May)

Heavily grazed plant (May)

Leaves and berries in July

Shoot in October

Lightly grazed stems in winter

Cowberry *Vaccinium vitis-idaea*

A **short** dwarf shrub (mostly <25 cm tall) with **oval, untoothed, leathery, dark, evergreen leaves** 1-3 cm long, small pale pinkish flowers in summer and **red berries** in autumn. Similar to Bearberry (page 64) but more upright and with leaves **down-turned along their edges and lacking the pattern of dark veins on their undersides**. Grows as **scattered plants** among other species or forming **loose patches**. On **dry to moist, acid soils** in **dry heaths** (H10 & H12-22), **birchwoods** (W17) and **pinewoods** (W18); also in some **bogs** (M19). Found mainly in **colder** northern, high altitude or exposed places. It is relatively **unpalatable** to grazing animals such as sheep and deer.

Shoot with flowers

Shoot with fruits

Bog Bilberry *V. uliginosum* ► is rather similar in growth form and leaf shape but leaves **deciduous, paler greyish-green** (dull pinkish in autumn), many of them widest beyond the middle. Uncommon; mainly in Scottish Highland **montane heaths** (H13-22) and **bogs** (M19).

Bog Bilberry

Bearberry *Arctostaphylos uva-ursi*

A **low, creeping, mat-forming** dwarf shrub with **oval, untoothed, leathery, dark, evergreen leaves** 1–3 cm long, small pale pinkish flowers in summer and **red berries** in autumn. It resembles Cowberry (page 63) but is more **creeping** and its leaves are **not down-turned along their edges** and have a clear pattern of **pale veins visible on their upper surface and dark veins on the otherwise pale lower surface**. Mainly on **dry to moist, acid soils** in well-lit places: in **dry heaths** (especially H16 but also H10 & H12–17) and less commonly wet heaths (M15) and bogs (M17–19). Also on basic soil; for example on limestone and basalt. It grows best **where competition from other plants is reduced** by **rockiness**, by **exposure to wind**, or by **burning**. It **can be abundant for some years after moor-burning** until taken over by taller, thicker Heather. It is rare outside the Scottish Highlands.

Leaves in March

Leaves and flowers in May

Arctic Bearberry *A. alpinus* ► has creeping shoots forming small, low patches, and **slightly larger leaves** that can be widest beyond the middle, have **toothed edges**, **very distinctively wrinkled upper surfaces** and turn **bright red in autumn**. In montane **heaths** (esp. H17 but also H13–15/19/20) and **bogs** (in some M19) in the Scottish Highlands and Orkney/Shetland.

Arctic Bearberry in October

Bog-myrtle *Myrica gale*

A small, branched shrub with **oval, grey-green, deciduous leaves** up to 6 cm long, **widest beyond the middle** and commonly with a few small teeth near the tip. Differs from small willows in its **red-brown twigs and buds**, and **strongly smelling buds and leaves**. Small **catkins** (5-15 mm long) in spring: male orange-brown; female greenish. Grows on **damp to wet, peaty soils** in **wet heaths** (M15-16), **bogs** (M17, 18 & 21) and **Purple Moor-grass mires** (M25). **At least moderately palatable** to grazing animals; can grow 2 m tall in ungrazed places; grazed plants generally <50 cm tall. Flowers commonest on less grazed plants.

Dwarf Birch *Betula nana* ► is a shrub <1 m tall with **short-stalked round leaves with relatively large teeth**. Scarce in **upland bog** (M19) & **wet heath** (M15) in the Scottish Highlands; very rare in N Engl. **Palatable** to grazing animals.

Creeping Willow *Salix repens*

A low shrub, mostly <30 cm tall and creeping, but can reach 1.5 m. **Narrow oval untoothed greyish-green leaves with silvery hairy undersides** (when young, uppersides hairy too) and down-turned edges. Young twigs are hairy. Flowers are catkins, in spring. In **dune slacks** (SD13-17), **heaths** (esp. H7 & M15-16) and **grasslands** (incl. MC10) on **damp to dry, acid to neutral soils**; commonest near coasts. Rare in fens.

See also **Dwarf Willow** *Salix herbacea* on page 370.

Common Restharrow *Ononis repens*

A low shrub up to 60 cm tall, with **hairy stems** (on which spines are present on some plants) and hairy, toothed, **greyish-green, scented leaves with 3 leaflets** (up to 2 cm long) and a pair of leaf-like stipules at base of leaf stalk. Pale **pink flowers** in summer. On **dry basic soils** in the **lowlands**, in **grasslands** (esp. CG3-5, CG7) and **sand dunes** (SD7-19).

Spiny Restharrow *O. spinosa* is similar but **more spiny** and with **hairs on stems in 1-2 lines running up stem** (hairs are all around stem in Common Restharrow). In **grassland** (esp. CG3-5) on **basic soils**, and in **coastal marshes**.In lowlands of England (except extreme SW), NE & SE Wales, and Solway Firth coast of Scotland.

Cranberry *Vaccinium oxycoccos*

Unmistakeable with its **thin, creeping wiry stems** (0.5–0.7 mm wide) with **small oblong-oval or triangular-oval, hairless leaves** 5–8 mm long, their **edges untoothed and curled down**. Leaves rather leathery-textured and persisting through winter (when they may turn dark brown). **Pink flowers** in summer; **edible red berries** in autumn. Plants trail low over *Sphagnum* mosses and other low-grown plants in **acid bogs** (esp. M18 raised bogs). Widespread, but rare in N Scotland and SE half of England. **Small Cranberry** *V. microcarpum* is smaller, with stems 0.3 mm wide and leaves 3–7 mm long; in similar habitats in Scotland and (v. rare) NE England.

Bog-rosemary *Andromeda polifolia*

Another plant found among *Sphagnum* carpets in **acid bogs** (M18) commonly with Cranberry. The **short, upright shoots** bear **narrow, dark green, hairless leaves** up to 3.5 cm long, with distinct **pale net-veined patterns**. **Pink flowers** in summer. Mainly at low altitudes in Wales, N England, SW Scotland and central Ireland.

Wild Thyme *Thymus drucei*

Thyme could easily be taken for a herb but the **slender, square-sectioned, trailing, patch-forming stems** are **woody**. They bear **small, oval, blunt-tipped evergreen leaves 4-8 mm long** in **opposite pairs**; leaves can be **hairy**. In summer there are **small pink flowers in short, blunt spikes**. Thyme grows mainly on **thin, dry, basic soils**, in **short or open, grazed calcareous grasslands** (CG1-2 & CG7-12) and **dwarf shrub heaths** (H7-8, H10 & H20, & CG13-14). It also grows more sparsely in taller, less grazed calcareous grasslands (CG3-6), and on cliffs. It is a **good indicator of basic soils**. In grasslands it **can decline if grazing is reduced** and the vegetation becomes tall and dense, with a build-up of a deep lower layer of moss, plant litter and humus.

Plant in flower in June

Plant in January, with dead flower heads from previous year

Large Thyme *T. pulegioides* is larger and more upright, with **hairless leaves 6-11 mm long**. In **dry lowland calcareous grassland** (CG2/3/7) in **south and east England**.

Rockrose *Helianthemum nummularium*

This is another low-grown, prettily-flowering dwarf shrub that could easily be taken for a herb, but the slender stems are woody in their older, lower parts. The leaves are like those of Thyme (with which it commonly grows) in being **small, oval, blunt-tipped, evergreen** and in **opposite pairs**, but they are larger (**5–20 mm long**), with a **prominent central vein in a groove on the upperside**, and a **whitish underside**. They can be **hairy**. There are **two stipules** (like tiny narrow pointed leaves) **at each leaf base**; Thyme lacks these. Large **yellow flowers** are conspicuous in summer. Rockrose grows on **thin, dry, neutral to basic soils**, in **short, grazed grasslands** (CG1–10 & U4c) and in very open vegetation on **steep rocky banks**. A useful **indicator of basic soils**.

Flowering plant in June

Closer view of leaves in September

Raspberry *Rubus idaeus*

Rather bramble-like, but stems more upright (to 1.5 m; upper parts of stems commonly curved), **rounded in section and not climbing or scrambling, much less prickly** (numerous short, thin, straightish prickles <2 mm long on stems and leaf stalks), and **leaves brighter green above, whitish-green beneath and not evergreen**. Small greenish-white flowers in summer, later becoming **red edible raspberries**. On **dry to moist, acid to neutral soils**, mostly in or at the edges of **woods** (esp. W10-11) and among **bramble and bracken** (W24-25) in woodland glades or at woodland edges. **Palatable** to animals such as sheep and deer: height and condition are **useful indicators of grazing pressure**.

Tutsan *Hypericum androsaemum*

A **well-branched** plant up to 1 m tall, with **unstalked, oval, untoothed, blunt-tipped leaves 5-10 cm long**. **Yellow, 5-petalled flowers** in summer, in branched groups, later becoming **oval, berry-like fruits** that are **red** at first but then **black**. It is related to St John's-worts and is included with them on page 244. The combination of berries and opposite oval leaves is reminiscent of Honeysuckle (page 76), but that species is more of a scrambler or climber, its flowers are cream-coloured and its berries are in denser clusters and do not turn black. Tutsan grows mainly on **dry or moist, acid to basic soils** in **woods** (esp. in W8-11) in S & W Britain & Ireland. It also grows on sea cliffs. It is **widespread in eastern Britain as an introduction**.

Close-up of flower

Plant in July with flowers and (upper right) last year's berries

Young shoot in April

Plant in December, with black berries

Section 1d

CLIMBERS & SCRAMBLERS

These are plants whose stems scramble or climb up trees, shrubs, rocks, banks and walls. They are most common in woods and hedges. When growing up trees and shrubs they do not parasitize their host plants but merely use them for support.

Climbers and scramblers are generally palatable to grazing animals such as deer, sheep and cattle, so they grow most abundantly and luxuriantly where grazing is no more than light: in some such places arboreal climbers can become a significant element of the total woodland structure.

Most of these species have woody stems. Five species on page 78 actually have non-woody stems but are described here so as not to split up a distinctive group of tall and conspicuous hedgerow scramblers and climbers.

Some non-woody plants described elsewhere in this book can also have a scrambling or climbing habit: certain Vetches (page 183), Fumitories (page 184, including Climbing Corydalis), Bedstraws (page 213) and Greater Bird's-foot-trefoil (page 195).

Bramble *Rubus fruticosus* agg.*

The **long, arching, red-tinged stems** of Bramble are **angled** in section and bear **strong, wide-based, straightish or backward-curved prickles**. The long-stalked **evergreen leaves** are **divided into 5-7 leaflets in a radiating pattern**, with **prickles on their stalks and on the undersides of their veins**. The white or pale pink, 5-petalled flowers become **edible blackberries** in late summer and autumn. Grows mainly on **dry to moist, acid to basic soils** at low altitudes, in **woods** (esp. W8-14), **scrub** (W21-23), among **bracken** (W25) and **forming thickets** (W24) along **hedges, field margins** and on **disturbed or waste ground**. **Palatable** to grazing animals: most abundant and luxuriant where there is **little or no grazing**.

* *agg. = an aggregate of more than 320 microspecies.*

Roses (see page 74) have stems more rounded in section and deciduous leaves with large stipules (like little leaves) on each side of the lower portion of the leaf stalk. Raspberry (page 70) is not scrambling or climbing, has non-angled stems with thinner, weaker prickles, and its leaves are deciduous and have very pale undersides.

Plant in September

Berries in September

Stem in April (note angled edges)

Flowers in June

Dewberry *R. caesius* is similar but **stems round in section** and **stems and berries with bluish 'bloom'** as on sloes, damsons, grapes etc. On **dry basic soils** in **grassland, dunes, scrub, hedges and woodland edges** in lowlands N to Firth of Forth.

Roses *Rosa* species

Roses have a **climbing habit** and **prickly stems** as in bramble (page 73) but the **stems are rounded in section** so they look smoother (apart from the prickles!). Their **leaves are not evergreen** and are **divided into 5–7 leaflets**, which, apart from the end one, are **in two rows along the leaf stalk. Large stipules (leaf-like structures) are attached to each side of the lower part of the leaf stalk. Large white or pink flowers** in summer, becoming **red 'hips'** in autumn. On **dry to moist soils**, especially in lowland **woods and hedgerows.** The commonest species are:

Dog-rose *R. canina*: **hairless green** upper leaf surface; prickles all about equal size, stout and **hooked**, with bases as wide as prickles are long; stems tough and upright or arching.

Field Rose *R. arvensis*: **hairless grey-green** upper leaf surface; prickles all about equal size, slender and **hooked**, with bases not as wide as prickles are long; many slender arching/trailing stems; flower has small projecting column-like structure (c.5 mm long with widened tip) in centre; widespread but rare in W Ireland & N of central Scotland.

Sweet-briar *R. rubiginosa*: **hairless dark green** upper leaf surface (but hairs present on undersides of veins); **leaf underside apple-scented**; prickles of unequal size, **hooked**, with bases as wide as the prickles are long.

Burnet Rose *R. spinosissima*: **hairless grey-green** leaves; **straight prickles**.

Sherard's Downy-rose *R. sherardii*: **blue-green** upper leaf surface; **dense short hairs** on both surfaces of leaf; **leaf underside resin-scented**; prickles slender and many of them **hooked**, with bases up to about as wide as the prickles are long; branches commonly twisted.

Soft Downy-rose *R. mollis*: like *R. sherardii* but **branches and prickles straight**.

Sherard's Downy-rose: flower in June (L); fruit (hip) in September (R)

Dog-rose stem and leaves in June

Ivy *Hedera helix*

Easily identified by the **dark, glossy, evergreen, long-stalked leaves** that have **pale veins** on their uppersides and are **oval or with 3-5 lobes** in a radiating pattern. Mainly on **dry to moist, neutral to basic soils** in **woodland** (especially the lowland types W8, W10 & W12) and **scrub** (W21-22). Also as a **climber on trees, shrubs and walls**, and on **cliffs** and among **screes**. Can form extensive ground carpets and lush growths up trees, especially where there has been some **artificial disturbance followed by years of neglect, with little or no grazing**; many such places are near human habitations. In more natural habitats mainly in **steep and rocky** places. **Not parasitic; does not damage host trees**, but dense growths **can smother tree-dwelling mosses, liverworts & lichens** (a matter of concern where ivy grows among rich moss, liverwort and lichen assemblages on trees and rocks). Very vigorous populations in tree canopies can overshade (and outcompete) tree foliage, and their weight can even bring down weak trees. Ivy is **palatable** to grazing animals.

Ground carpet of ivy leaves (May)

Lobed leaves on ground (January)

Unlobed leaves with flowers, on a tree (January)

Honeysuckle *Lonicera periclymenum*

Distinctive **woody, twining stems** with **oval, short-stalked or unstalked leaves in opposite pairs**. Unmistakeable **creamy flowers** in summer, later becoming clusters of **red berries**. Leaves emerge in early spring and fall off in winter. Stems creep over ground, and climb up trees and shrubs, and also up walls and steep rocky banks. Stems can twist around each other. On **dry to moist, acid to neutral soils**, mainly in **woods** (esp. W10–11) and **hedges**. Commonest where **lightly grazed or ungrazed**, as it is **palatable** to grazing animals such as sheep, deer and cattle. Luxuriant growths up trees indicate that grazing is, or at least has been, no more than light for several years.

Climbing stems (May)

Plant in May

Plant in October

Flower in September

Fruit in September

Traveller's-joy *Clematis vitalba*

Also called **Old Man's Beard** because of the **conspicuous masses of long, whitish 'whiskers' projecting from the fruits in autumn-winter**. A **climber** with **woody, ridged stems**, and **leaves** that are in **opposite pairs** and are **divided into 5 stalked, oval leaflets** that can have a few teeth. Larger leaflets can have heart-shaped bases. Greenish-white, 4-petalled flowers (2 cm across) in summer. Leaves fall off in winter. **Stems climb up or scramble over trees, shrubs, cliffs, walls and banks.** Grows on **dry, basic soils**, in **lightly grazed or ungrazed hedges**, **scrub** (W21) and **woods** (W8), and on **cliffs, walls** and **railway banks**. **A good indicator of basic soils.** Native in **lowlands of southern Britain**. **Introduced further north** (but scarce N of Cumbria) **and in Ireland.**

Plants in October, with conspicuous long whitish hairy 'plumes' on fruits

Other tall climbers of hedges and woods

Six species found **climbing and scrambling up and among shrubs and trees in hedges, woodland, scrub** and **fens**, mainly on **dry to moist neutral to basic soils in the lowlands**. They can also grow on **disturbed or waste ground**.

Hop *Humulus lupulus* has **leaves with 3 or 5 radiating lobes with toothed edges**. Leaves persist through winter. Small white male flowers, in clusters. **Female flowers have green scales** and become **brown fruits in autumn**. Stems **non-woody**. In **lowlands of England & Wales N to Lancashire/Yorkshire**; **introduced elsewhere**.

White Bryony *Bryonia dioica* has **leaves with 5 radiating lobes with untoothed edges**. Whitish flowers 12–18 mm across; **red berries in autumn**. Stems **non-woody**. Native in **English lowlands** (but not SW England); rare in Wales; **introduced elsewhere** in lowlands.

Hedge Bindweed *Calystegia sepium* has **heart-shaped leaves**. **Large white flowers** 4–7 cm across (summer). No berries. Stems **non-woody**. **Widespread in lowlands**.

Large Bindweed *C. silvatica* is similar to Hedge Bindweed but the leaf base curves around into the leaf stalk (this junction is abrupt in Hedge Bindweed). **Introduced**.

Bittersweet *Solanum dulcamara* has **triangular or heart-shaped leaves** that **can have small lobes at their bases**. **Drooping purple & yellow flowers** (with petals curled back) in clusters in summer. **Red berries** in autumn. Stems are **woody**. **Widespread in lowlands**.

Black Bryony *Tamus communis* has **shiny, heart-shaped shiny leaves with sunken veins**. **Small white flowers** 4–5 mm across, becoming **red berries** in autumn. Stems **non-woody**. In **lowlands of southern half of Britain** (N to S Cumbria and Co. Durham).

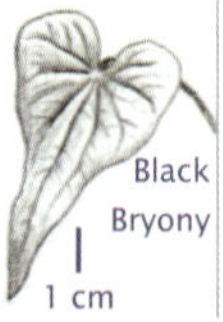

Hop (October)

White Bryony (June)

Hedge Bindweed (October)

Bittersweet (June)

Bittersweet flowers (June)

Section 2

GRASSES AND OTHER GRASS-LIKE PLANTS

These have long, narrow leaves and rather inconspicuous green or brown flowers that lack petals. They have adapted successfully to a wide range of environments and form much of the vegetation in most types of habitat in Britain and Ireland. For their identification it helps to understand the division of leaves into **sheaths, ligules** and **blades**, and (in grasses) to know what **nodes** are. These features are illustrated on the next page and referred to in the table below. Note that the cereal crops wheat, barley, oats and maize (sweetcorn) are grasses (see page 118).

Easily observed differences between grasses, sedges, rushes and woodrushes

	Stems with swollen nodes?	Stems hollow or solid?	Shape of stem section	Leaf ligule present?	Leaf blade present?
GRASSES	Yes	Hollow	Round	Yes (not fused to blade) [2]	Yes
SEDGES					
Most species	No	Solid	3-sided [1]	Yes (fused to blade) [2]	Yes
Deergrass	No	Solid	Round	No	Yes [3]
Spike-rushes	No	Solid	Round	No	No
RUSHES	No	Mostly solid [4]	Round	No	Yes/No [5]
WOOD-RUSHES	No	Solid	Round	No	Yes [6]

1 3-sided in most of these species but round in some.
2 In sedges, the ligule is for most of its length stuck (fused) to the upper surface of the base of the leaf blade. In grasses the ligule is 'free': not fused to the leaf blade.
3 The leaf blade of Deergrass is just a very short flap.
4 Some species have inside of stem filled with soft 'pith'.
5 Very varied: some species have long leaf blades but others none at all.
6 Woodrush leaf blades are quite wide and flat (like those of many grasses) but are distinct in having long white hairs along their edges.

Section 2a

GRASSES

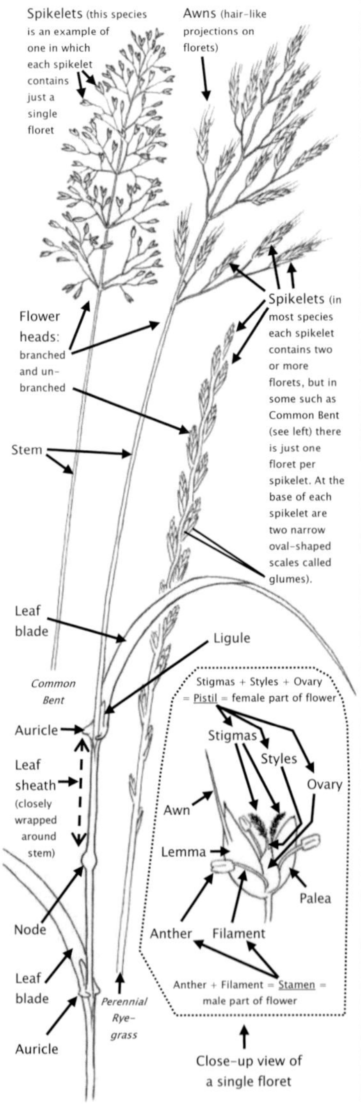

The drawings on this page show the main aerial (above ground) parts of a grass. Two of the drawings are of particular species and are named accordingly; the third (the middle one) is a composite, with features of more than one species. Main grass features:

- **LEAF BLADE**: the uppermost part of the leaf; the sticking-out part.
- **LEAF SHEATH**: the lowest part of the leaf, wrapped closely around the stem.
- **LIGULE**: a small flap where the top of the leaf sheath joins the base of the leaf blade.
- **AURICLES**: little leafy projections around the top of the leaf sheath: they are absent in many species.
- **NODES**: swollen points at intervals along the stem; the bottom of the leaf sheath starts at a node.
- The individual flower of a grass is a **FLORET**, and these are grouped together into larger **SPIKELETS** (though in Bent Grasses each spikelet contains just a single floret). Some species have **AWNS**, which are very thin, hair-like projections arising from the florets; they can be conspicuous (for example in barley).

Grasses differ from sedges, rushes and woodrushes in having **hollow stems**, swollen **nodes** on these stems, and in their **ligules** being **not fused** to the upper surface of the base of the leaf blade (sedge ligules are fused).

For convenience, a drawing showing typical grass leaf and stem features is repeated on each page on which a grass species is described.

Some growth forms and leaf shapes of grasses

These photos and drawings illustrate certain growth forms and leaf and ligule shapes of grasses. These shapes are referred to in parts of the text.

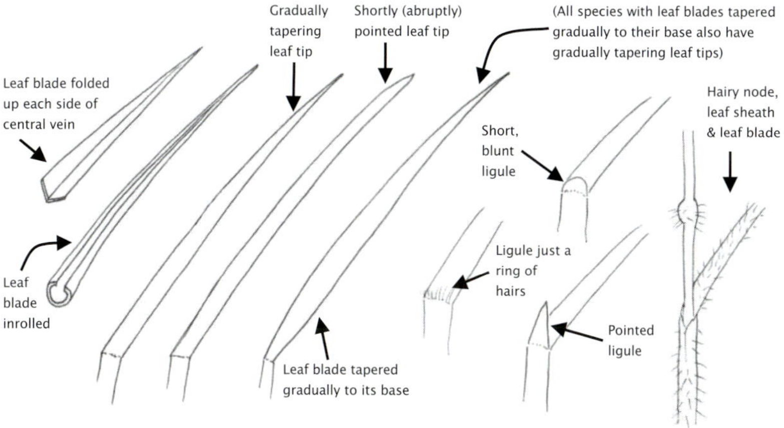

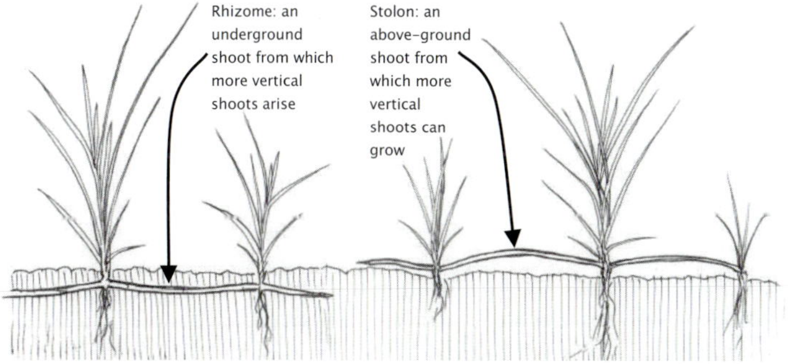

Dense tussock (Tufted Hair-grass)

Small, thin tuft (Wavy Hair-grass)

Grass species grouped by flower head shape

The grasses in this book are split into those with **very thin, wiry leaves** <1 mm wide (pages 84–88) and those with **wider leaves** (pages 89–122). If plants are flowering, you can speed up identification by looking at the **general shape of the flower head** (see table below); otherwise, see table of leaf features on next page.

THIN, UNBRANCHED flower head with SPIKELETS UNSTALKED so they grow directly off the main stem (or on extremely short stalks visible only on such close examination that the flower head still looks unbranched) 	WIDER, BRANCHED flower head whose SPIKELETS ARE ON STALKS			
	REGULAR, SYMMETRICAL flower head with branches in groups (whorls) at regular intervals along the stem 		More IRREGULARLY-BRANCHED and LESS SYMMETRICAL flower head	
	FLOWER HEAD SMALL (mostly <15 cm long) on STEMS MOSTLY <50 CM TALL	FLOWER HEAD LARGE (mostly >20 cm long); STEM TALL (50–200cm)	DROOPING SPIKELETS on long, widely-spaced branches 	SPIKELETS NOT DROOPING  
Mat Grass, Sheep's & Fine-leaved Sheep's Fescues, Early Hair-grass, **Meadow & Marsh Foxtails, Timothy, Smaller Cat's-tail, Couch, False Brome, Tor-grass, Crested Hair-grass, Blue Moor-grass, Marram, Perennial Rye-grass, Sweet Vernal-grass, Crested Dog's-tail, Wheat, Barley, LYME-GRASS**	Bristle Bent, **Common Bent, Velvet Bent, Brown Bent, Creeping Bent, Smooth, Spreading, Rough and Wood Meadow-grasses, Yellow Oat-grass**	**Tufted Hair-grass, Wood Millet, COMMON REED, MAIZE**	**Hairy-brome, Giant Fescue, Barren Brome, Quaking Grass, Oats**	Wavy Hair-grass, Sheep's, Fine-leaved Sheep's, Viviparous & Red Fescues, Common Saltmarsh Grass, **Cocksfoot, False, Meadow & Hairy Oat-grasses, Purple Moor-grass, Soft & Upright Bromes, Wood Melick, Yorkshire-fog, Creeping Soft-Grass, Annual Meadow-grass, REED CANARY-GRASS**
Species names in non-bold type = leaves very thin and wiry (<1 mm wide); **bold** = wider leaves (1–14 mm wide); **BOLD UPPER CASE** = leaves very wide (at least some leaves >15 mm wide).				

Summary table of leaf features of grasses

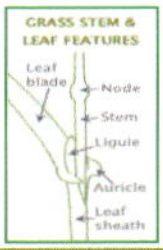

(T) in left column = species that can form dense tussocks.
+ (in left column) = incl. 3 subspecies: Smooth, Spreading & Narrow-leaved Meadow-grasses.
* in the 'Leaf tip...' column = leaf tapers gradually to its base as well as its tip.

Species (ordered by leaf width: narrow-leaved species first)	Max. leaf blade width (mm)	Max. leaf blade length (cm)	Max. ligule length (mm)	Tapering leaf tip: gradual or abrupt	Leaf blade hairy or greyish? ( ) = hairy or greyish on some leaves/plants	Leaf sheath hairy?	Auricles?	PAGE
Mat Grass (T)	1	20	2	Gradual				85
Wavy Hair-grass (T)	1	20	3	Either				86
Bristle Bent T)	1	20	5	Gradual				86
Sheep's Fescue/Early hair-gr.	1	20/5	<1	Either				87
Red Fescue (lower leaves)	1	40	<1	Abrupt				88
Crested Hair-grass (T)	2.5	20	0.7	Abrupt	(Hairy)	(Hairy)		90
Velvet Bent	2.5	15	4	Gradual				95
Brown Bent (T)	3	15	4	Gradual				95
Common Saltmarsh Grass	3	20	3	Abrupt	Greyish			91
Red Fescue (stem leaves)	3	20	<0.5	Gradual	(Greyish)			88
Wood Meadow-grass (T)	3	12	1.3	Either				96
Quaking Grass	4	15	1	Gradual				92
Crested Dog's-tail	4	15	2	Gradual				93
Yellow Oat-grass	4	15	1	Gradual	Hairy	Hairy		94
Common Bent	4	15	2	Gradual				95
Heath Grass (T)	4	25	0	Abrupt	(Hairy) (Gr)	Hairy		98
Smooth Meadow-grass(T)+	4	30	2	Abrupt	(Greyish)			96
Rough Meadow-grass	4.5	20	10	Abrupt				96
Annual Meadow-grass	5	14	5	Abrupt				97
Smaller Cat's-tail	5	12	3	Gradual				112
Sweet Vernal-grass	5	30	5	Gradual	(Hairy)	(Hairy)		99
Hairy oat-grass	5	30	5	Abrupt	Hairy	Hairy		101
Meadow Oat-grass (T)	5	30	5	Abrupt	Greyish			101
Tufted Hair-grass (T)	5	60	14	Either				100
Blue Moor-grass (T)	6	20	0.7	Abrupt	Greyish			102
Perennial Rye-Grass	6	20	1.5	Abrupt			Y	103
Upright Brome (T)	6	30	3	Gradual	(Hairy)	(Hairy)		105
Marram (T)	6	60	30	Gradual	Greyish			120
Wood Melick	7	20	1	Gradual	Hairy	Hairy		106
Marsh Foxtail	7	12	5	Gradual				104
Soft Brome	7	20	2.5	Gradual	Hairy	Hairy		107
Barren Brome	7	25	4	Gradual	Hairy	Hairy		107
Creeping Bent	8	10	6	Gradual				95
Tor-grass (T)	8	45	2.5	Grad. (some=*)	(Hairy)	(Hairy)		116
Timothy	8	45	6	Gradual				112
Meadow Fescue	8	45	1	Gradual			Y	113
Yorkshire-fog	10	20	4	Gradual	Hairy (Gr)	Hairy(Gr)		108
Couch	10	30	1	Gradual			Y	111
Wheat	10	30	2	Gradual	Greyish		Y	118
Meadow Foxtail	10	40	2.5	Gradual				112
False Oat-grass (T)	10	40	3	Gradual				114
Purple Moor-grass (T)	10	45	0	Gradual*				110
Tall Fescue (T)	10	60	1.6	Gradual			Y	113
Creeping Soft-Grass	12	20	4.5	Gradual	(Hairy)	(Hairy)		109
False Brome (T)	12	35	5	Gradual*	Hairy	Hairy		115
Cocksfoot (T)	14	45	9	Gradual				117
Barley	15	30	1.5	Gradual	Greyish		Y	118
Oats	15	45	5	Gradual				118
Giant Fescue	15	60	2	Gradual			Y	119
Wood Millet	15	30	10	Gradual				119
Hairy-brome	16	60	4.5	Gradual	Hairy	Hairy		119
Reed Canary-grass	18	35	16	Gradual				121
Lyme-grass	20	60	1	Gradual	Greyish			120
Common Reed	30	60	0	Gradual*	Greyish			122
Maize	100	100	3	Gradual	(Hairy)	(Hairy)		118

Wiry-leaved grasses (pages 85-88)

These are grasses whose leaves are so thin (<1 mm) and wiry that they look the same width all round (from side to side and top to bottom), so it is not immediately clear which is the upper surface and which is the lower.

There are fewer wiry-leaved grass species than broader-leaved ones. This helps with identification: for example, a wiry-leaved grass growing in woodland is most likely to be Wavy Hair-grass, and one that grows as pale, very dense, coarse tussocks on the hills is most likely to be Mat Grass.

One species - Red Fescue - also has some broader leaves but these are only on the flowering stems and are less numerous and less conspicuous than the very thin, wiry leaves of the plant bases and non-flowering shoots.

Some wiry-leaved grasses can look rather like the three wiry-leaved sedges described in this book: Hare's-tail Cottongrass, Deergrass and Black Bog-rush. However, those species differ as follows: Hare's-tail Cottongrass (page 138) has white cottony fruiting heads, a 3-angled leaf section, and blades of stem leaves arising from very inflated sheaths; the wiry 'leaves' of Deergrass (page 140) are actually stems with small flower heads at their tips, its true leaves being tiny and inconspicuous on the lower parts of stems; Black Bog-rush (page 141) has very dark, dense flower heads and, at the shoot bases, shiny blackish leaf sheaths. Heath Rush (page 145) might also be mistaken for a wiry-leaved grass but is easily told by its leaves being so unusually stiff and glossy, forming low and remarkably dense rosettes; the leaves are also channelled and about 2 mm wide.

Wiry leaves: these ones belong to Red Fescue

Mat Grass *Nardus stricta*

This distinctive grass forms **dense tufts** of **stiff wiry leaves** whose blades **stick out at a wide angle from their sheaths**. Flower heads are **narrow, dark and one-sided**. The whole plant is **very pale in winter**. Grows mainly on **moist acid** upland soils. **Unpalatable** to grazing animals such as sheep, deer and cattle. Commonest in **upland grasslands** (U5 & U7) where growth of competitors is reduced by **heavy grazing** or by a **cold** climate as on high slopes where snow lies well into the spring.

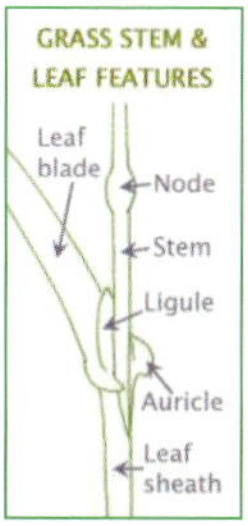

Plant in late May, with distinctive dark, narrow flower spikes (inset = single flower head)

Leaf blades stick out at wide angle from sheaths and stems

Nardus grassland looking very pale in early spring; known as 'White Moor' in some areas

Wavy Hair-grass *Avenella flexuosa*

This has tufts of **wiry leaves** as in fescues but with **purplish sheaths** (green in fescues), **ligules up to 3 mm long** (shorter in fescues) and **delicate flower heads** with **small 2-flowered shiny purple-tinged spikelets on wavy stalks**. Tufts are small and narrow or larger and dense. On **dry to moist, acid soils** in **woods** (esp. W16/17), **grasslands** (U2-7), **heaths** and **mountain habitats** (many NVC types), **bogs** (esp. M20) and **felled conifer plantations**. Appears **not very palatable** to animals. In woods it is the commonest wiry-leaved grass and can form dense swards where **grazing** limits dwarf shrub growth. *D. flexuosa* grassland (U2) is usually a temporary result of **disturbance** (**burning** or **tree-felling**): after some years other plants tend to take over.

Leaves and flower head in woodland in July

Bristle Bent *Agrostis curtisii* ▸ looks similar but has **dense tufts of straight, rather greyish-green leaves**, **narrow flower spikes** with **straight (not wavy) branches**, and a **longer ligule (up to 5 mm)**. On **dry to moist, acid soils**, in **heaths** (mainly H3-4, esp. where **burned**) and as **grassland** swards (U3), in **SW England** (E to Sussex) & **extreme S Wales**.

Sheep's Fescue *Festuca ovina*

This species forms **small, low tufts** of **wiry leaves** with **very short (barely visible) ligules**, stems 10-25 cm tall and **narrow flower heads** mostly <7 cm long, with pointed spikelets 5-7 mm long. Compared with Red Fescue it has **shorter leaves** <15 cm long (lower leaves up to 40 cm in Red F.), **shorter stems** <30 cm long, **smaller, narrower flower heads** with spikelets very shortly-stalked (longer-stalked in Red F.), and **leaf sheaths with left & right sides joined together only in lower part** (joined all way up in Red F.). Grows on **dry to moist, acid to basic soils** in **short, grazed, unimproved grasslands** (U1-6; CG1-2; CG7-13), **heaths and mountain habitats** (H1-22; U7-17), under **bracken** (U20) and **among rocks**. **Moderately palatable** to grazing animals.

Plants flowering in June, with some older, paler flower heads in the photo on the right

THREE SIMILAR SPECIES: **Fine-leaved Sheep's Fescue** *F. filiformis* has **shorter (4.5-5.5 mm) spikelets**; in **dry to moist acid grassland**. **Viviparous Fescue** *F vivipara* has **viviparous** flower heads (lots of 'baby plants'). **Dry to moist upland grasslands** (U4-7, U13 & CG10-13) and various **mountain habitats** (U8-17). **Early Hair-grass** *Aira praecox* is **shorter** (stems mostly <15 cm; leaves <5 cm) with **very small flower heads** (<2 cm long) whose **spikelets have projecting hairs 3-4 mm long** (<2 mm in Sheep's Fescue). Flowers in spring. On **thin, dry, acid soils**, esp. in **open or rocky lowland grassland** (U1).

Viviparous Fescue

Early Hair-grass (August) in short acid grassland (U1)

Red Fescue *Festuca rubra*

Most leaves wiry (see photo on page 84) but **those on stems are up to 3 mm wide**. Forms **patches**: not very tufted. Less tufted than Wavy Hair-grass, with **wider stem leaves, shorter (barely visible) ligule,** and **flower head with larger spikelets** (oval & pointed with hair-like projections (awns)) **on straight stalks.** Told from Sheep's Fescue by the **longer basal (lower) leaves** (up to 40 cm; <15 cm in Sheep's Fescue), **broader stem leaves, longer stem** (mostly >30 cm; <30 cm in Sheep's Fescue), **wider, more open flower head** (lower branches longer and stick out more than in Sheep's Fescue) and **leaf sheaths with left & right sides joined right up to the top** (joined only to halfway up in Sheep's Fescue). Mainly on **dry to moist, neutral soils**, in **unimproved grassland** (MG1-5) and on **sea cliffs** (MC1-12). Also on **acid** soils (U4), **basic** soils (CG1-3 & 6), on **sand dunes** (SD5-19) and in **saltmarshes** (esp. SM16-21, 24-25 & 28). Commonest at low altitudes. **Moderately palatable** to grazing animals.

Plants in July and (inset) June

Wider-leaved grasses (pages 90-122)

These are grasses with most leaf blades more than 2 mm wide: this is wide enough to have distinct upper and lower surfaces and left and right edges.

This is a large and varied group of species, among which there are many similarities especially when plants are not in flower. But don't be put off! Many species can be identified using easily observed features such as growth form (tussocky or not), the width of the leaves, the leaf texture (hairy or hairless, stiff or soft) and colour (bright green, grey-green etc), the shape of the leaf tip (gradually or abruptly pointed), the length and shape of the ligule, and whether or not the leaf blades taper to their bases.

Winter and early spring are generally the hardest times of year for identification of these grasses, because the previous year's growth has had plenty of time to die away and the next year's growth has not, or has only barely, begun. However, some useful identification features can persist through the winter. Of course some species fare better than others in this respect. For example the leaves of Tufted Hair-grass are so tough that their unmistakeable features are always obvious. Sometimes we are defeated by something too old and decayed, or too young and tiny, or just too 'scrappy'. It is unrealistic to expect that every single grass specimen should be easily identifiable in the field regardless of its size or state or the time of year, but it is always worth a try and I hope this section of the book makes the identification process more straightforward and interesting.

The table on page 83should help to narrow things down using features of the leaves. The descriptions of species (pages 90-122) are ordered mainly by leaf width (beginning with those with narrower leaves) but with some species moved slightly so that certain pairs or groups of species with other important things in common can be put on the same page: for example all the Bent Grasses (closely related; similar flower heads and general appearance), and also Lyme-grass and Marram (big grey-green grasses of sand dunes).

If flowering heads are present, so much the better - more clues to narrowing things down at an early stage (see the table on page 82).

False Oat-grass in February

Crested Hair-grass *Koeleria macrantha*

Short, tufted and rather hairy. The leaf blade is up to 2.5 mm wide and 20 cm long, has a **shortly pointed tip** and a 'furrowed' upper surface with about six **narrow ridges running parallel** and very close together along the length of the leaf. The **flower head is narrow and dense** like that of Sheep's Fescue (but thicker) or Sweet Vernal-grass (but smaller, not glossy and with smaller individual florets). Grows mainly in short, **grazed, unimproved grassland** (CG1-10 & 13, & U1) on **thin, dry, calcareous soils**, especially in the lowlands. **Appears moderately palatable** to grazing animals. Tufted Hair-grass has similar furrows on the leaves but is much bigger, with a branched flower head on a tall stem (to 1 m), its leaves are much longer (to 60 cm), stiffer and very rough when felt backwards, and it grows mainly on damp soils.

L: Flowering heads in June. R: Variation in degree of open-ness of flowering heads in June

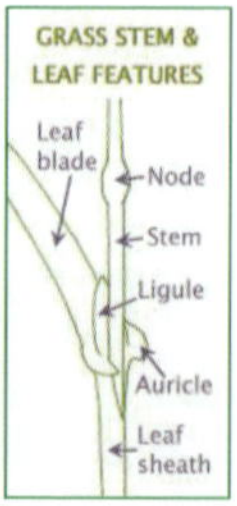

Leaves at base of plant, in June

Common Saltmarsh Grass *Puccinellia maritima*

This grass forms **low, dense mats** of **creeping or slightly ascending shoots** of a distinctive appearance: the **wide leaf sheaths** make the **shoots look thick**, but the **leaf blades are short and thin**, and stick out **alternately to left and right**. The leaf blades can be up to 3 mm wide but commonly have their edges folded upwards along each side, making them look narrower. They have **blunt tips**. The flower head is branched, with rather long, blunt-ended spikelets, but is narrow and inconspicuous. Common on **saltmarshes**.

Leafy shoots in September, showing alternating left-right arrangement of leaves. Also present are Sea Aster leaves (wider) at left, and Glasswort (behind, at upper right)

Flower head (a relatively narrow, little-branched one) in September

Quaking Grass *Briza media*

Unmistakeable flower heads with **large, flattened, triangular spikelets hanging on long, fine branches**. Each spikelet has many florets. Leaf hairless, up to 4 mm wide and tapering to a fine point, and with a very short (0.5-1 mm), blunt ligule, as in Common Bent and Crested Dog's-tail. Leaf sheath very tight-fitting (looser in Common Bent). Crested Dog's-tail leaves differ in having shiny undersides, and young leaves folded along the central vein. Mainly in **unimproved, grazed or mown short grassland on dry to moist, neutral to calcareous soils** (MG5 and CG1-6 & 8-10). Also in **herb-rich mires** (M22-24 & 26). Appears to be **at least moderately palatable** to grazing animals.

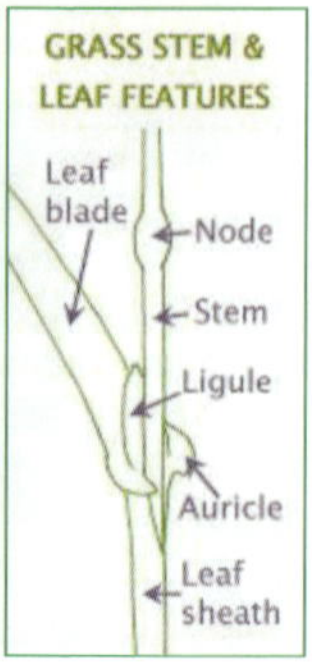

Plants in June-July

Crested Dog's-tail *Cynosurus cristatus*

Most easily told by the **distinctive narrow, unbranched flower head** rather like that of wheat but much smaller and thinner. The flower heads can persist through winter, which helps with identification at that time of year. The **leaf is hairless**, up to 4 mm wide, **tapering gradually to a fine tip** and with a **short, blunt ligule** 0.5–2 mm long. In these respects it resembles Common Bent, but has a more **shiny underside**, a tighter-fitting sheath and the **young leaves are folded up along each side of the central vein**. The leaves are like those of Perennial Rye-grass but more gradually tapering, and the lower leaf sheaths are not purplish. Grows mainly in **grazed or mown, unimproved or semi-improved short grasslands on dry to moist neutral soils** (MG3–6).

Flowering stems in mid July

Yellow Oat-grass *Trisetum flavescens*

When present in summer the **flower head** is distinctive: **rather long** (5-15 cm) and **much branched** as in Bent Grasses and Meadow-grasses but **glossy yellow-green** and with **longer spikelets each with long hair-like projections** (awns) up to 9 mm long. Flowering stems up to 80 cm tall. The leaves are **yellow-green with hairy upper surfaces**; the **sheaths of most lower leaves are hairy too**. Leaf blades are 2-4 mm wide.The **ligule is distinctive: very short (up to 1 mm) with many tiny, fine teeth along the otherwise blunt tip**. Soft-brome and Upright Brome can have similar ligules but their leaves are duller and not yellow-green (hairier and greyer in Soft-brome; with long marginal hairs in Upright Brome). Crested Hair-grass has a very short ligule (but without fine teeth) and hairy leaves, but its upper leaf surfaces are distinctly furrowed. Sweet Vernal-grass leaves can be similar but with a longer ligule and long hairs at the top of the leaf sheath. The leaves of Yorkshire-fog are greyer; those of Hairy Oat-grass are more abruptly pointed and those of False Brome are wider and gradually tapering to a narrowed base. Yellow Oat-grass grows on **dry or dryish, neutral to basic soils in the lowlands**, mainly in medium-height to tall **unimproved grasslands** (MG1-5 & CG1-8) **but also in some agriculturally improved grasslands** (MG6). It is **palatable** to grazing animals.

Flower head in July

Leaf blade & sheath in July

Bent Grasses *Agrostis* species

Grasses forming small tufts with tapering, finely-pointed leaf blades mostly <4 mm wide and stems to about 40 cm tall. Leaves as in Sweet Vernal-grass (but no long hairs at top of sheath), Crested Dog's-tail (but with non-shiny undersides, looser-fitting sheaths, and young leaves not folded along central vein) and Quaking grass (but with looser sheath). Symmetrical, branched flower head unlike those of other grasses just mentioned: regular branches point up or spread out, and bear many tiny 1-flowered spikelets. Generally palatable to grazing animals.

Common Bent *A. capillaris*. Blunt, untoothed ligule <2 mm long. Leaf blades mostly flat (to 4 mm wide), but edges can curl upwards. On dry to moist, acid to slightly basic soils, in unimproved, grazed short grasslands (U1-6, MG3-5 & CG10-13), under bracken (U20) and in woods (especially W11).

Velvet Bent *A. canina*. Ligule 2-4 mm long, mostly with untoothed, pointed or blunt tip. Leaf blade up to 3 mm wide; rather soft-textured; flat or with edges curled up. Tufts not dense; linked by creeping above-ground shoots (stolons - easily seen in open/patchy vegetation but less so in dense swards) rooting at their nodes. Mainly on damp acid soils, in heaths, mires and unimproved grasslands.

Brown Bent *A. vinealis*. Ligule 1-4 mm long; mostly with toothed/ragged, pointed or blunt tip. Leaf blade up to 2.5 mm wide; firm-textured; flat or with edges curled upwards. Densely tufted. No above-ground creeping stolons. Mainly on dry acid soils in heaths, woods & unimproved grasslands.

Creeping Bent *A. stolonifera*. Blunt ligule 1-6 mm long. Leaf blades flat; up to 8 mm wide. Tufts linked by above-ground creeping shoots (stolons) rooting at nodes. Mainly on dry to wet neutral soils in grasslands, periodically inundated habitats (including saltmarshes) and disturbed habitats. Leaves like those of Marsh Foxtail (for difference see bottom of p. 104).

Water Bent *Polypogon viridis* looks like Creeping Bent but has a wide jagged ligule tip and the flowers fall off after flowering, leaving bare branches. Introduced. In disturbed/urban habitats.

Creeping Bent flowering shoot

Bristle Bent *A. curtisii* is very different: dense tufts of thin wiry grey-green leaves; in S England & S Wales; more details on page 86.

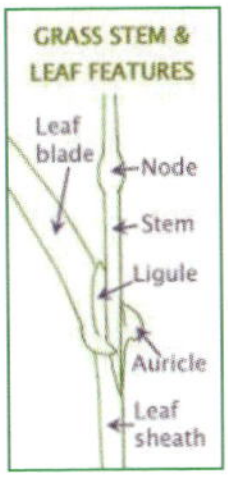

Creeping Bent leaves

Meadow-grasses *Poa* species

(see also Annual Meadow-grass on page 97)

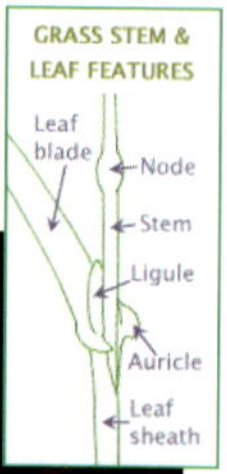

The species on this page are mostly 20-50 cm tall and **more or less hairless**, with **leaf blades parallel-sided, 2-4 mm wide and with 'sudden' shortly-pointed tips**. Other species with leaves of similar shape and size are Perennial Rye-grass (whose leaf blade has auricles at base and distinct parallel ridges/grooves on upperside), Meadow Oat-grass and Blue Moor-grass (dense tufts of strongly blue-grey-tinged leaves), Hairy oat-grass (leaves hairy), Crested Hair-grass (leaf has furrowed upper surface and can be hairy) and Common Saltmarsh-grass (wide leaf sheaths; saltmarsh habitat). The **symmetrical flower heads have small spikelets on spreading branches** as in Bent Grasses (page 95), but with larger spikelets each with 3 or more florets (1 floret in Bent Grasses).

Rough Meadow-grass (July)

Smooth/Spreading/Narrow-leaved Meadow-grass *Poa pratensis s.l.* * grows up to 60 cm tall, but commonly much shorter. **Smooth leaf sheaths** and a **blunt ligule 1-2 mm long**. **Smooth M-g.** *P. pratensis*: as described; very common **Spreading M-g.** *P. humilis*: similar but **less tufted** and **can have greyer leaves**; very common. **Narrow-leaved M-g.** *P. angustifolia* has **lower leaves narrow with inrolled edges**; uncommon outside SE half of England. These three species can be hard to separate. Mostly in **short grassland** on **dry to moist, neutral, basic or slightly acid soils** (U4, MG1-7 & CG1-10); also **sand dunes**, **disturbed/waste ground** and (esp. *P. humilis/angustifolia*) the **tops of walls**. Appears to be **palatable** to grazing animals.

Rough Meadow-grass *P. trivialis* has **rough leaf sheaths** and a **pointed ligule 4-10 mm long**. Forms **loose tufts joined by above-ground creeping shoots** (stolons). Mostly on **moist neutral soils** in **improved or unimproved grasslands** (esp. MG1-3 & 5-10), **woods** (esp. W7-9) and **hedgerows**. Appears **palatable** to grazing animals.

Wood Meadow-grass *P. nemoralis* is **tufted** and hairless, has **smooth leaf sheaths**, a **blunt ligule <1.5 mm long** and narrow leaf blades <3 mm wide and **abruptly or gradually pointed**. Leaves characteristically **stick out more or less horizontally from the erect stems**, so massed stems and their leaves **look like miniature scaffolding**. Flower spikelets can be very small. Whole plant has a **delicate appearance**. Mostly on **dryish, more or less neutral soils** in **woods** (esp. W8-11 & W14) and **hedges**, and also among mountain rocks. Native, but **probably introduced in Ireland & NW Scotland**. **Palatability** (to grazing animals) unclear.

* = *sensu lato* = in the broad sense: *P. pratensis, P. humilis* and *P. angustifolia* together.

Annual Meadow-grass *Poa annua*

A **very short, tufted, hairless** grass with **flattened leaf sheaths (the lower ones pale whitish-green)** and a **blunt ligule 2–5 mm long.** (Cocksfoot also has flattened leaf sheaths but is much bigger.) **Leaf blades can be flattened and folded, are short (mostly <10 cm long) and up to 5 mm wide, with abruptly-pointed tips. Young leaf blades can be crinkled. Small, branched, pale green flower heads**, each spikelet with 3–5 florets. Common on **dry or moist neutral soils** where vegetation is sufficiently short or sparse for small plants to survive without being smothered by larger ones: in **gardens**, along **roadsides, paths, hedgerows** and **field margins**, and on **disturbed or waste ground**. **Tolerant of trampling** by people and farm animals.

Plants in April

Heath Grass *Danthonia decumbens*

This is the only grass in this book that has **abruptly pointed leaf tips combined with the absence of a ligule**. There is just a line of very short hairs where a ligule would be, and, more conspicuously, at each end of this line there are **long, conspicuous whitish hairs sticking out to each side at the top of the leaf sheath** (which itself is hairy). The leaves grow in dense but low tufts. Leaf blades are up to 4 mm wide and can reach 25 cm in length but are commonly much shorter. They can be hairless or hairy, but the sheaths are more consistently hairy. The leaves can also be greyish, resembling those of Spreading Meadow-grass, Meadow Oat-grass and Blue Moor-grass, but all of those species have a distinct ligule and no white hairs at the top of the leaf sheath. The flower heads (on stems typically 20–40 cm tall) are quite small and have **relatively few but large spikelets**, and are greenish or with a purple tinge (see photo). Heath Grass grows on **dry to damp, acid to basic soils**, in **short, grazed, unimproved grasslands** (U3-5, MG5 & CG1-14) and **heaths** (H3-8, 10, 12, 16, 18 & M15-16). It is most common in upland areas. It appears to be **at least moderately palatable** to grazing animals such as sheep, cattle and deer.

Plant in June

Sweet Vernal-grass *Anthoxanthum odoratum*

Easily told by the distinctive **unbranched oval flower spike** whose long, shiny, pointed spikelets become very pale in late summer, and the **long hairs at the extreme top of the leaf sheath** (a feature also seen in Heath Grass, but that species lacks a ligule and has abruptly pointed leaf tips). The leaf blade is up to 5 mm wide, **tapers gradually** to a fine pointed tip (as in many other species), and **tastes of almonds**. The ligule is 1–5 mm long. Where it grows with Common Bent or Creeping Soft-grass the leaves look brighter green and the plant a bit more tufted. On **dry to moist, acid to neutral soils**; mainly in **unimproved short grasslands** (especially U4 & CG10–13), **woodland** (especially W11), but also under **bracken** (U20) and in some damp habitats such as rush mires (M23). **Palatable** to grazing animals such as sheep, cattle and deer; generally in **grazed** vegetation (otherwise outcompeted by taller grasses).

Flower heads in mid May...

...and mid July

Long hairs at extreme top of leaf sheath (mid May)

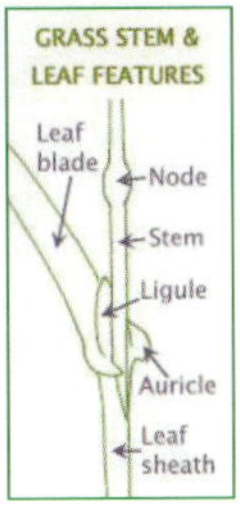

Tufted Hair-grass *Deschampsia cespitosa*

Dense tufts of **long, stiff, dark green, parallel-sided leaves** up to 5 mm wide, whose upper surfaces have **parallel ridges and grooves** running their length and are **very rough** if felt in a backwards direction from tip towards base of leaf (roughness caused by tiny forward-pointing hairs). **Large, branched flower heads** with many small, well-spaced 2-flowered spikelets, on **tall** stems (60–100 cm). On **damp neutral soils,** in **grassland** (dominant in MG9), **woods** (mainly W7–9) and on **cliffs** (U17), and on **moist acid soils in mountain grasslands** (U13 – plants generally shorter here). **Moderately palatable**: can be grazed down to dense stubby rosettes in some heavily-grazed pastures, but being coarse and fibrous it is commonly left to grow large and long-leaved.

Above: flowering heads in August. Below: plants in April with last year's pale old leaves

Oat-grasses *Helictotrichon/Helictochloa*

Densely tufted grasses with distinctive flower heads up to 18 cm long (on stems up to 60 cm tall) with **relatively few long, narrow spikelets** (mixed silver & dark brown in colour) **with long projecting 'hairs'** (awns). **Leaf blades up to 5 mm wide**. Found mainly in **lightly to heavily-grazed, unimproved, dry calcareous grassland** (CG1-10), especially in the lowlands. These grasses appear to be **moderately palatable** to grazing animals such as sheep and deer.

Meadow Oat-grass *Helictochloa pratensis* is **hairless**, with a **stiff leaf blade** with a **shortly pointed tip** and a **blue-green** upper surface (a waxy 'bloom' that can be rubbed off with a fingertip), and a **ligule 3-5 mm long**. The flower stem is quite stiff. Blue Moor-grass (page 102) has a similar leaf but differs in its shorter ligule and small, dense flower head. Found mainly in **lightly to heavily-grazed, unimproved, dry calcareous grassland** (CG1-10), especially in the lowlands.

Hairy Oat-grass *Helictotrichon pubescens* is similar but with **softer-textured leaves that lack a bluish bloom and have hairy sheaths**; mainly in **unimproved grasslands** on **dry to moist, neutral to calcareous soils** (esp. MG1-3, MG5, CG2-3, CG6-8).

Meadow Oat-grass in June

Blue Moor-grass *Sesleria caerulea*

A **densely tufted, hairless** grass with **leaves up to 6 mm wide, shortly pointed at their tips** and with **blue-green upper surfaces**. It differs from Meadow Oat-grass (which shares the above features) in its **extremely short ligule** and the **small, dense, oval flower head** (long and open in Meadow Oat-grass). Spreading Meadow-grass has similar leaves but is not tufted and has a very different (larger, branched and more open) flower head. Blue Moor-grass is common in **dry, calcareous, unimproved, heavily to quite lightly grazed grassland** (CG8-9) on limestone in parts of N England and W Ireland, occurring in both upland and lowland areas. It also grows sparingly in upland grasslands and cliff ledges in the southern parts of the Scottish Highlands. It is rather **unpalatable** to grazing animals such as sheep and cattle.

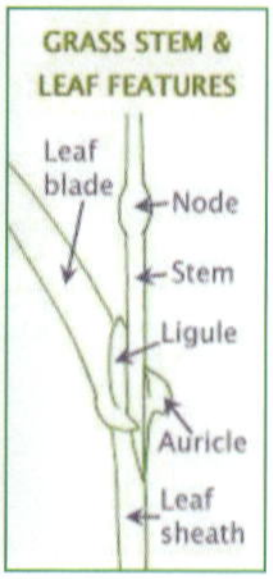

Plants in June, including close-up views of leaves and a flower head

Perennial Rye-Grass *Lolium perenne*

Unmistakable **long, narrow, flattened, shiny green flower heads** with small spikelets alternately left & right along a slightly zigzag stem. Stems and flower heads are tough and can persist through winter, when pale brown. **Leaf blades hairless, 2-6 mm wide and shortly-pointed**, with **small auricles** at their bases, many **parallel ridges/grooves on their uppersides**, and **shiny undersides**. **Purplish leaf sheaths at base of plant**. Young leaf blades flattened (folded either side of central vein). Forms tufts or patches. Common in agriculturally **improved or reseeded neutral grassland** (MG6-7), **semi-improved grasslands** (MG3a/4/5a & U4b) and on **disturbed or waste ground** (OV20-23). **Palatable** to grazing animals and **tolerant of grazing and trampling**.

Flower heads in June

Plants in July

Lower stems and leaves in June

Italian Rye-grass *L. multiflorum* is similar but with **leaves mostly 6-9 mm wide** and flower spikelets with **projecting hair-like awns**. **Introduced**; mainly on **disturbed ground in lowlands**.

Marsh Foxtail *Alopecurus geniculatus*

A **short** grass whose stems **creep** in their lower parts, then turn upwards, **bending at each node**, to become upright with **dense, narrow flower heads** like those of Meadow Foxtail (page 112) but smaller and darker. **Leaf sheaths are wide and wrapped loose around stems** so the shoots look quite thick. **Leaf blades are up to 7 mm wide and relatively short**: up to 12 cm long. The ligule is blunt and 2-5 mm long. Grows on **moist or wet neutral soils**, in **short, grazed or mown, unimproved or partly improved grassland** (MG13) and **by water**; especially in places that are **flooded from time to time**. Commonly with Creeping Bent (page 95) whose leaves are similar but with shallower ridges on their uppersides - see diagram below).

Flowering plants in May

Close-up showing stems bent at the nodes

Profiles of ridges/grooves on upper surface of leaf blades (seen through a hand lens)

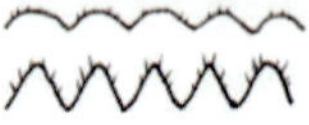

Agrostis stolonifera: grooves shallow; ridges wider than tall

Alopecurus geniculatus: grooves deep; ridges as tall as wide

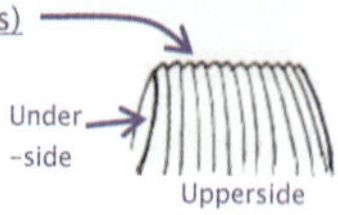

Upright Brome *Bromopsis erecta*

This grass forms **dense tufts** of **tall stiff stems** with **hairless or slightly hairy leaves**. The **lower leaves are thin (2–3 mm) and rolled like tubes**. The **leaves on the stems are wider (to 6 mm), flat and with long bristly hairs along their edges**. Ligules are up to 3 mm long. The **large, branched flower heads** have several long (15–35 mm) multi-flowered spikelets. Two rather similar grasses are Red Fescue (page 88), which is smaller and shorter, with narrower leaves and a shorter ligule, and Tall Fescue (page 113), which has all leaf blades wide and flat, without marginal hairs and with distinct auricles. Upright Brome is locally common in the **south** in **lightly grazed** **chalk & limestone grassland** (CG3 & 5). It is **palatable** to grazing animals.

Hedgerow plants in June

Close-up of leaf blade, showing long hairs along its edges

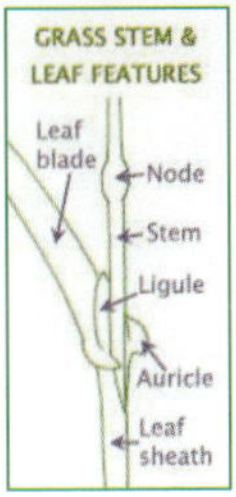

Wood Melick *Melica uniflora*

Distinctive with its **bright green leaves** that are rather hairy, soft-textured and 3-7 mm wide. Bright colour and tendency for some leaves to droop at tips can give resemblance to False-brome (page 115), but: leaves shorter, narrower, less hairy and not widening out in mid-leaf; plants form **patches** (not tufts); flower heads are branched with **well-spaced, oval, bluntish, brown spikelets**; **uniquely among British & Irish grasses, there is a pointed ligule-like spike** 1-4 mm long **at the top of the leaf sheath but on the opposite side of the stem to the leaf blade** (a true ligule is on the same side as the leaf blade). Grows on **dry to moist, neutral to basic soils** in **woods** (esp. W8-11) and **hedgerows**, especially in the lowlands.

A patch of Wood Melick in May

Leaf in April

Flower heads in May and (inset) June

Soft-brome *Bromus hordeaceus*

This species grows up to about 60 cm tall. Its **leaves are grey-green and softly hairy**, with blades up to 7 mm wide. In this respect it resembles Yorkshire-fog (page 108), but the **ligule is shorter** (to 2.5 mm compared with up to 4 mm in Yorkshire-fog) **with small teeth along its broad, blunt tip** (blunt and untoothed in Y-f.). The **flower head** is very different: a **loosely branched group of large** (1-2 cm long) **spikelets** that are **covered in short hairs** and also have **longer, projecting, hair-like awns**. Mainly on **dry, neutral soils at low altitudes**, especially in **disturbed but little-grazed and untrampled places** such as **field margins, hedgerows, meadows, tracksides** and **waste ground**.

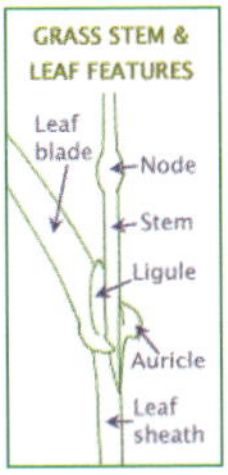

< Plants in June

Barren Brome *Anisantha sterilis*

The stems and **softly hairy leaves** of this species are similar to those of Soft-brome and Yorkshire-fog, but the **ligule (to 4 mm long) has many small teeth along its pointed (not blunt) tip**. The **flowering heads are large, with long spikelets on drooping, widely-spreading branches**. The spikelets have many **long, hair-like projections** (awns). Mainly on **dry, neutral soils in the lowlands**, in **disturbed but little-grazed and little-trampled** habitats such as **roadsides, hedgerows, field margins** and **waste ground**. Rare in the Highlands and in western Ireland.

Plant in June

Flower head like that of Hairy-brome (which is taller, with auricles, blunt untoothed ligule and stiffer stem) and Giant Fescue (hairless, with blunt untoothed ligule).

Yorkshire-fog *Holcus lanatus*

A tufted grass with **soft-textured, hairy leaves and stems**. Leaf blades are up to 10 mm wide and **greyish tinged**. Leaf **sheaths are commonly tinged pink** (many narrow vertical pink-purple stripes visible at close range). The pale grey-green to pale pinkish, soft-textured flower head varies from narrow (closed) to triangular (open with spreading branches). Grows mainly on **dry to moist, neutral soils** in short to tall **grasslands** (esp. MG1 & MG3-10, U4, MC8-9 & MC12, SD8-9 & SD13-17), **woodland** (esp. W10) and weedy vegetation in **disturbed** habitats (mainly OV19/23/25/27). **Palatable** to grazing animals. Where abundant it can indicate some **agricultural treatment** (e.g. nitrogen enrichment). Creeping soft-grass is similar but less tufted, its leaves are less hairy, its stems hairless with whitish tufts at the nodes, and with a hair-like projection (awn) up to 5 mm long on each floret in the flower head. The hairs on Yorkshire fog leaf sheaths can point downwards, as in Hairy-brome (taller, stiffer, rougher-textured, with stouter stem and much bigger flower head) and Wood False-brome (yellower green leaf blades gradually narrowed to their bases and very different flower head with two rows of long narrow spikelets).

Plants in July

Lower stems and leaf sheaths in May

Young Creeping Soft-grass with stems + hairy nodes not yet apparent looks like Yorkshire Fog, but the tiny hairs on the ligules are all very short (YF has a mix of short and longer hairs).

Creeping Soft-Grass *Holcus mollis*

Identifiable by the softly hairy, **soft-textured leaves** 3-12 mm wide and up to 20 cm long, the stems hairless except for conspicuous tufts of **white hairs at the nodes** ('hairy knees'), and leaf sheaths mostly lacking the pink-purple stripes seen in Yorkshire Fog. Flower head as in Yorkshire-fog (soft-textured and varies from narrow to wide triangular), but less pink-tinged and **each floret with a hair-like projection (awn) 3.5-5 mm long**. **Not very tufted**; spreads by underground creeping shoots (rhizomes) from which stems grow up; can form **patches**. Leaves can grow well up the stems, as in Common Reed, Reed Canary-grass (both with taller thicker stems and wider leaves) and Couch (which lacks hairy nodes). False Oat-grass can have hairy nodes but is tufted, has taller, thicker stems, longer leaves (to 45 cm) and larger, shiny flower-heads. Creeping Soft-grass grows mainly on **dry to moist, neutral to acid** soils in **woodland** (W10-11), but can also be common in open **neutral grassland**. Appears **moderately palatable** to grazing animals.

Dense sward in August

Stems in May

Young plants with stems + hairy nodes not yet apparent can resemble Yorkshire Fog, especially if they also happen to have pink-tinged leaf sheaths, but looking through a lens the tiny hairs on the ligules are seen to be all very short (Yorkshire Fog has a mix of short and longer hairs).

Purple Moor-grass *Molinia caerulea*

Tall and tufted, with hairless or slightly hairy leaves whose **blades are up to 45 cm long and 3-10 mm wide, gradually narrowed toward both tip and base**. The leaves turn **pale gold** in autumn and **fall off in winter** to accumulate as **mats of pale, curly leaf litter**. The leaf blades commonly droop at their tips. There is **no proper ligule: just a ring of hairs**. Plants can form **tussocks** raised on 'stools' hidden by the long leaves. Non-tufted young/small plants look very different but still have the telling 'ring-of-hairs' ligule and leaf blade narrowed at base (as in Common Reed, which has much wider leaves). The flower head is long, narrow and branched, with many small spikelets. Purple Moor-grass grows mainly on **moist or wet, acidic, peaty** soils, especially with some lateral water flow. It is dominant in some **grasslands/mires** (M24-M26) and is very common in **wet heaths** (H2-6 & M15-16), **bogs** (M17-18 & M21) and wet **woods** (W4). *Molinia* grassland is very extensive in western upland areas; much of it is the result of previous **burning** and/or **grazing** where the vegetation would otherwise be heathy. Young leaves can have **some grazing value in spring and early summer** but soon become fibrous and **unpalatable**, but can be grazed by some cattle breeds right through into autumn.

Sward in May (young & old leaves visible)

Tussock in October

Leaves and old flowering stems in October

Tussocks in winter

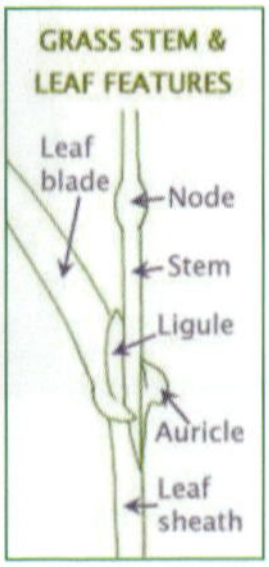

Common Couch *Elymus repens*

A rather tall grass with **hairless or very slightly hairy leaves up to about 10 mm wide**. Not very tufted; forms dense **patches**. Ligules are up to **1 mm** long and there are tiny **auricles**. Leaves can look like those of False Oat-grass and Tall Fescue (both of which are tufted), and Timothy and Meadow Foxtail (which lack auricles and have dark brown lower old leaf sheaths, Timothy also with a longer ligule up to 6 mm). **Leaves can be plentiful a long way up the stems**, as in Common Reed and Reed Canary-grass (both bigger, with much wider leaves). The **narrow flower head has two rows of spikelets arranged alternately left and right** as in Perennial Rye-grass (page 103) but the spikelets stick out more (see photo below). Grows on **dry to moist, neutral soils** in **lowlands**, in **ungrazed or lightly grazed grassland** (MG1), along **hedgerows** and **field margins**, on **disturbed (but not very trampled or grazed) and waste ground** and the drier parts of **saltmarshes** (SM28). Spreads by underground creeping shoots (rhizomes); can be a troublesome **garden weed**.

Leafy stem in August

Flower head in July

Bearded Couch *E. caninus*: similar but **spikelets have projecting hair-like awns up to 2 cm long**. Flower head 'nods' to one side. In **woods** (esp. W7-10) **on neutral/calcareous soils**.

Sea Couch *E. athericus*: like Common Couch but **blue-green**; **leaves inrolled at edges**. On **coastal saltmarshes** (SM24). Rare in Scotland (Solway coast). Illustration at top of page 79.

Sand Couch *E. junceiformis*: also **blue-green** but **shorter**; **flower spikelets well spaced and not overlapping**. On **sand** (less commonly shingle) **at seaward edge of dunes** (SD4 - shown in lower-right of upper photo on page 376).

Meadow Foxtail *Alopecurus pratensis*
Timothy *Phleum pratense*

Tall, tufted grasses resembling False Oat-grass (page 114), Couch (page 111) and Tall Fescue (page 113) in their **long, hairless or very slightly hairy leaves up to 45 cm long and about 10 mm wide**, but with **narrow, dense, cylindrical flower spikes** and with the **lowest, oldest leaf sheaths dark brown**. Leaves lack auricles (unlike Couch & Tall Fescue). Mainly on **dry to moist, neutral soils** in **lowlands**, in **ungrazed or lightly grazed grassland** (MG1), and along **hedgerows** and **field margins.**

Meadow Foxtail *Alopecurus pratensis* has a **cigar-shaped flower head**, the individual (very closely-spaced) florets each with a hair-like projection (awn) 3-6 mm long, giving the flower head a soft appearance. Ligule up to **2.5 mm** long. In **grasslands, hedges and field margins**; also in some **reseeded grass swards** (MG7).

Timothy *Phleum pratense* has a **rougher, blunt, cylindrical flower head** with shorter hair-like projections <2 mm long. Ligule up to **6 mm** long: longer than in Meadow Foxtail, Couch or False Oat-grass. In **grasslands, hedges and field margins**, and some **reseeded grass swards** (MG7).

Meadow Foxtail flower heads

Timothy flower heads

Smaller Cat's-tail *Phleum bertolonii* is like Timothy but shorter (<70 cm) and smaller, with **leaf blade <5 mm wide** and **<12 cm long, ligule <3 mm long** and **flower head <7 mm wide** (>7 mm in Timothy); in **short neutral to basic unimproved lowland grasslands**.

Tall Fescue *Schedonorus arundinaceus*

This **tufted, hairless grass can be 1.5-2.0 m tall**, with **leaves up to 60 cm long and 10 mm wide**. It looks like False Oat-grass (page 114) but with **distinct auricles** (with short hairs along their edges - see drawing) and **widely-branched flower heads with fewer, larger spikelets** that are not shiny and whose florets lack long projecting hair-points. Flower head as in Red Fescue (but leaves much wider), Upright Brome (but leaves wider and with auricles, and florets lacking long hair-points) and Meadow Fescue (but plant more densely tufted, leaves wider, and auricles bigger and hairy-edged). Grows on **dry to moist, neutral soils** in the **lowlands**, especially in **ungrazed or lightly grazed grassland** (MG1 & MG12) and along **hedgerows** and the **upper parts of saltmarshes. Palatable** to grazing animals.

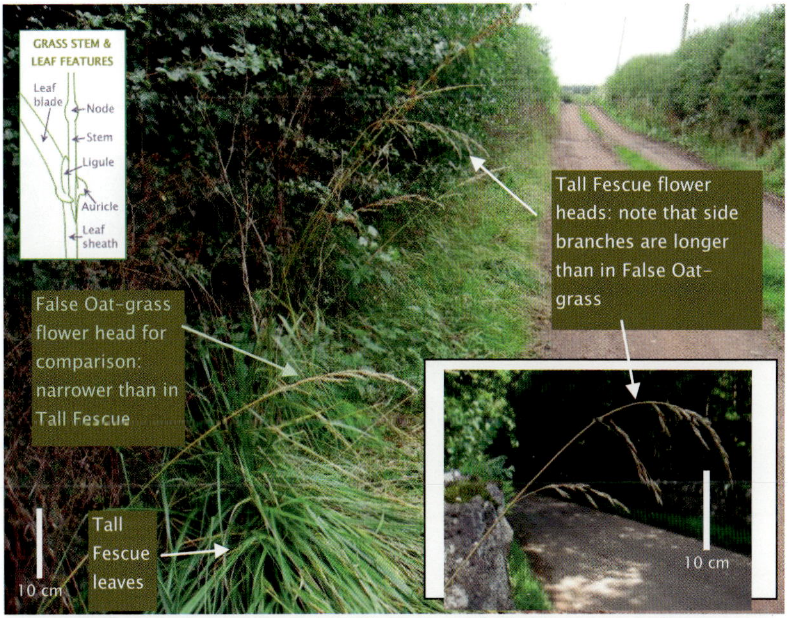

Hedgerow plant in August (with flower head of False Oat-grass in front)

Meadow Fescue *S. pratensis* is **shorter** (mainly <1 m), with **narrower leaves** (<8 mm wide) with **smaller auricles lacking marginal hairs**. Mainly in **grasslands** on **neutral soils** in **lowlands**; also in hedgerows and on waste ground.

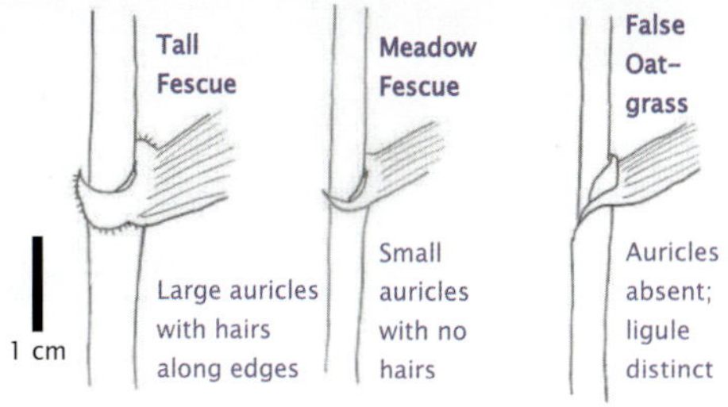

False Oat-grass *Arrhenatherum elatius*

Tall and tufted, with more or less hairless leaf blades up to 45 cm long and 4-10 mm wide. Ligules are 0.5-3 mm long. The **long, branched flower heads** have many narrow oval **shiny green** spikelets. The extreme stem base is typically orange-coloured. Grows on **dry to moist, neutral soils**, especially in **ungrazed or unmown lowland** places: **grassland** (MG1-2), **hedgerows**, woodland **glades**, **sand dunes** (SD9) and less commonly within **woodland** (esp. W8/10). **Palatable** to grazing animals. Can resemble Cocksfoot (which has flattened shoot bases and leaf sheaths, stiffer stems and 'blobby' flower heads), Tall Fescue (which has auricles), Couch, Timothy and Meadow Foxtail (which have narrow, unbranched flower heads; Couch also has auricles and the other two have old lower leaf sheaths dark brown; Timothy also has a longer ligule), and Creeping Soft-grass (which is not tufted, has a shorter, more slender stem, shorter leaves to 20 cm long and smaller, non-shiny flower-heads).

False Brome *Brachypodium sylvaticum*

A medium-sized, **tufted** grass 30-80 cm tall, with **bright green, hairy, soft-textured leaf blades** up to 35 cm long and 12 mm wide, **narrowed at their bases** and **drooping at their tips**. The flower head has 4-12 **long narrow unstalked spikelets** arranged alternately left and right along the stem. Grows on **dry to moist, neutral to basic soils**, in **woods** (W7-9), **hedgerows** and, less commonly, grasslands. Can resemble Tor-grass (page 116) whose leaves are not droop-tipped and leaf hairs are shorter, and Hairy-brome, which is taller and stiffer, with wider, branched flower heads and leaf blades not narrowed toward their bases.

Long white hairs on leaf sheath & blade (June)

Tor-grass *Brachypodium pinnatum s.l.*

A **tufted grass up to 1 m tall**, with **bright green leaf blades** up to about 45 cm long and 8 mm wide, the widest ones widening out in mid leaf from narrower bases as in False Brome (page 115) but typical leaves more parallel-sided. Leaf blades more **erect** than in False Brome; not drooping at tips. Leaf sheaths and blades **hairless or with very short hairs**: the hair length is no more than ¼ width of the rolled-up leaf sheath; False Brome has longer hairs whose length is at least ¾ width of the rolled-up leaf sheath. The flower head has **long narrow unstalked spikelets** arranged alternately left and right along the stem as in False Brome, but the projecting hair-points (awns) at the spikelet tips are no more than 5 mm long (up to 12 mm in False Brome). Tor-grass grows in **lightly grazed or ungrazed chalk & limestone grasslands** (CG4-5) in **southern Britain.** Two taxa, *B. pinnatum* and *B. rupestre*, are now recognized, but they are hard to tell apart - hence *B. pinnatum s.l.*.*

Leaves and flower head (June)

Can have very short hairs on leaf blade & sheath (June)

Habitat (NVC CG4) in June

* *s.l.* = *sensu lato* = in the broad sense: *B. pinnatum* and *B. rupestre* together.

Cocksfoot *Dactylis glomerata*

A **tall, tufted** grass with rough-textured, hairless, rather bluish-green leaf blades up to 45 cm long and 14 mm wide. **Bases of shoots are flattened (squashed-looking)**, with **leaf sheaths flattened** to give a fold line running up the back. The **stiff stem** has a green **flower head divided into dense 'blobby' clusters**. Flower heads of Reed Canary-grass (page 121) look quite similar but that species is taller, not tufted, has wider leaves and its shoot bases are not flattened. Cocksfoot grows on **dryish neutral soils**, especially in tall, **lightly grazed, ungrazed or unmown grassland** (MG1-2) in **fields, woodland glades, roadsides, hedges**, and **sand dunes** (SD9), **waste or disturbed ground** (OV23) and some ungrazed or little-grazed **woods** (esp. W10). Can survive as low, non-flowering rosettes in short, grazed grasslands.

Wheat, Barley, Oats & Maize (Sweetcorn)

Cereal crops grown on **dry, fertile soils in the lowlands**, especially in the drier east of Britain. They have broad leaves up to 15 mm wide in Wheat, Barley & Oats, and up to 10 cm wide in Maize. The first three turn pale golden in late summer when the oval-shaped grains in the flower heads are ripe and ready for harvesting.

Wheat *Triticum aestivum* has **thick, dense flower heads** whose many spikelets have **no projecting hairs or have just short inconspicuous hairs** (these hairs are actually called awns). **Used for animal food and flour.**

Barley *Hordeum distichon* is unmistakeable on account of the conspicuous and **extremely long, bristly hairs** (awns) projecting from the flower heads. **Used for animal food and for brewing beer and whisky.**

Oats *Avena* species have **branched flower heads** with large, drooping spikelets. **Used for animal food and to make oatmeal and porridge oats**.

Maize *Zea mays* is **up to 3 m tall**, with **leaves 3-10 cm wide**, stems 1-2 cm thick, and large, branched, brown flower heads. **Grown for sweetcorn**, mainly in S Britain.

Wheat in July

Barley in July

Oat in September

Barley field in July

Maize in September

Hairy-brome *Bromopsis ramosa*

Tall (to 1 m or more) and tufted, with long, downward-pointing bristly hairs on leaf sheaths. The **hairy leaf blades** are up to 60 cm long and 16 mm wide, with **auricles** at their bases. The **stiff stem** ends in a large flower head with **widely spaced spikelets on long, branched, drooping stalks**. On **dry to moist, neutral to basic** soils in **woods** (esp. in **ancient** W7-9). False Brome (page 115) can have downward pointing hairs on leaf sheaths but has shorter, narrower leaf blades (to 35 cm long & 12 mm wide) with narrowed bases, and unbranched flower heads. Yorkshire-fog (page 108) can have downward-pointing hairs on leaf sheaths but is shorter, with soft-textured leaves and a very different flower head.

Flowering head in June

Stems & leaves in June

Giant Fescue *Schedonorus giganteus* looks similar but is **hairless** and has **dark, purplish nodes** on the stems, and large, pale or mixed pale & dark **auricles**. It grows on **moist, neutral to basic** soils in **woods** (esp. in **ancient** W7-9).

Wood Millet *Milium effusum* is also tall with broad leaves up to 15 mm wide. Leaves **dull green** and **hairless**, without Giant Fescue's wide auricles. Branched flower heads with many **small, blunt, pale green spikelets** (a bit like flower heads of Meadow Grasses & Bent Grasses but bigger & more open). On **moist, neutral to basic** soil in **woods**, esp. **ancient** W7-9.

Lyme-grass *Leymus arenarius*

The only grass with **really wide (8–20 mm), tough, strongly blue/grey-tinged leaves** growing in **coastal sand**. Forms patches (SD5) at the **outer (seaward) edges of sand dunes**. Like Marram (see below) it spreads by means of underground creeping shoots (rhizomes) that help to stabilize loose sand.

Marram *Ammophila arenaria*

Dense tufts of long, coarse, sharp-tipped, greyish-green leaves that are **up to 6 mm wide but look thinner because their edges are rolled up** (like tubes). **Ligule very long: 1–3 cm**. Flower heads long and narrow (see photo). Common on **sand dunes** (SD6–9, SD11–12 & SD19). Spreads by means of underground creeping shoots (rhizomes) that help to stabilize loose sand.

Reed Canary-grass *Phalaris arundinacea*

A tall grass forming **patches of thick rattling stems up to 2 m tall** connected by underground creeping shoots (rhizomes). It is not tussocky. The **leaves are up to 18 mm wide** and **grow well up the stems** as in Common Reed (page 122). Differs from Common Reed in being shorter, with narrower leaves that are **not narrowed towards the base**, with a **ligule up to 16 mm long** (no ligule in Common Reed) and smaller, greener flower heads whose spikelets can form dense 'blobby' clusters as in Cocksfoot (which is tussocky, shorter and with narrower leaves and flattened shoot bases - see page 117). Reed Canary-grass grows on **damp or wet, neutral or eutrophicated soils** and in **shallow water**, forming **fens** (S28), and patches in **wet woods** (W1 & 6-7).

Plants in late June

Flower heads in July

Plants in winter

Plants in July

Common Reed *Phragmites australis*

A **very tall** grass forming **patches of thick rattling stems up to 3 m tall** connected by underground creeping shoots (rhizomes). It is not tussocky. The **leaves are very wide (to 3 cm) and grow further up the stems than in most grasses**. They are **gradually narrowed towards both tip and base**. There is **no proper ligule: just a ring of hairs**. The flower head is **large and purplish**: individual florets are very small. Common Reed grows on **wet, neutral, basic or eutrophicated soils** and in **shallow water**. It forms dense **swamps** (S4) and **fens** (S24-27) (these swamps & fens known collectively as Reedbeds), and also grows in **flushes and other mires** (M13, 15, 22-24 & 26-28) and **wet woods** (W2 & 7). More rarely it is found in bogs (M17, 18 & 21). Reed Canary-grass (page 121) looks similar but is smaller, with narrower leaves and a long ligule.

Reedbed (NVC S4) in August

Section 2b

SEDGES

These drawings show the main aerial (above ground) parts of a sedge. The larger drawing is a composite with features of more than one species; the other is Star Sedge.

Sedges differ from grasses in having **solid stems** (hollow in grasses) **without swollen nodes**, and their leaves with **ligules fused for most of their length to the upper surfaces of the base of the leaf blades**. The stems of many species are **3-sided** in section (grasses mostly round-sectioned) and the flower heads are of a different structure, many species having **separate male and female spikes** (male at stem tips; female lower down the stem). Most species have **leaves in three ranks**, sticking out in three directions around the stem, but this is not always obvious. Many species have **noticeably curved leaves with the central vein running in a distinct groove**: this makes them look different from typical grass leaves and gives them a little more rigidity too. Sedges grow in a wide range of habitat but are especially common on damper soils. Three species - Common Cottongrass, Hare's-tail Cottongrass and Deergrass - are among the commonest plant species in the extensive bogs and wet heaths of Britain and Ireland, and in such places they contribute significantly to the colour of the whole landscape.

This section begins with species with male spikes at the stem tips and more conspicuous female spikes further down the stem (pages 124-132), followed by those with all spikes similar in appearance and clustered near the stem tip (pages 133-141).

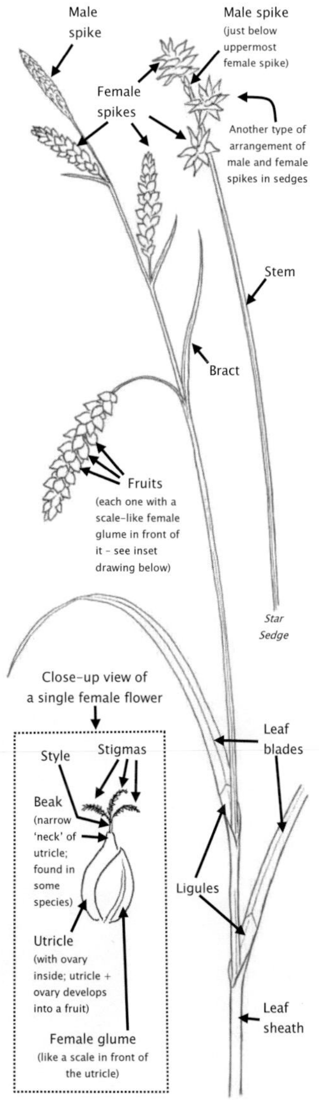

For convenience, a drawing showing the typical division of a sedge leaf into sheath, blade and ligule is repeated on each page on which these features are mentioned in the text.

Small to medium-sized sedges with grey-tinged leaves

These five species have leaves with a distinct **grey or blue-grey tinge**, though they can turn browner in autumn and winter. They grow mainly on damp or wet soils. Their flowering stems are mostly 20-40 cm tall, with leaves <30 cm long, but Bottle Sedge can be at least 60 cm tall and longer-leaved. These species are dealt with on four pages: some photos on this page, an identification table on the next page, followed by information on habitats, and finally more photos.

Carnation Sedge
(female spike in July)

Common Sedge
(female spikes in Sept.)

Glaucous Sedge
(female spike in April)

Carnation Sedge leaves seen from above (July)

Glaucous Sedge (June)

IDENTIFICATION OF SMALL TO MEDIUM-SIZED SEDGES WITH GREY-TINGED LEAVES

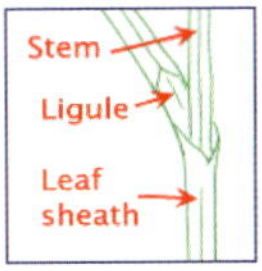

	* Basal sheaths	Ligule	Grey on leaves	Leaf size	Female spikes (+ no. of male spikes**)	Fruits
Common Sedge *Carex nigra*	Dark brown	Blunt	Grey-green on **both** sides.	**1-3 mm wide**; <30 cm long	**Unstalked**; **pointing up**; tapered base & tip; mottled **green/black**. *(1-2 male spikes)*	Small and oval. Numerous. **Densely crowded** together.
Glaucous Sedge *Carex flacca*	Dark brown	Blunt	Grey-green **under-side**; dull green upperside	**2-3 mm wide**; <30 cm long	**Stalked;** lower ones can **droop**; blunt; mottled **green/purple-brown**. *(1-3 male spikes)*	Oval to rounded. Numerous and **densely crowded** together.
Carnation Sedge *Carex panicea*	**Whitish or pinkish**	Blunt	Distinct (strong) pale blue-grey tinge on **both sides**.	**2-5 mm wide**; <25 cm long	**Short-stalked**; pointing up or out; blunt; mottled **green/purple-brown**. *(1 male)*	Roundish. Relatively few in no. **Loosely** arranged along spike.
Stiff Sedge *Carex bigelowii*	**Shiny red-brown**	**Point-ed**	Dull greyish-green on **both** sides.	**2-7 mm wide**; <20 cm long	**Unstalked** ; **pointing up**; bluntish; mottled **green/black**; clustered below single male spike at shoot tip. *(1)*	Small and oval. Numerous, and **densely crowded** together.
Bottle Sedge *Carex rostrata*	Dark brown	Blunt	Grey-green **upper-side**; dark green underside	**2-7 mm wide**; up to 1 m long	**Unstalked** or with v. short stalks; large; blunt; **yellow-green**. *(2-4 male spikes)*	Oval with projecting narrow tip. Numerous and **densely crowded**.

EXTRA NOTES TO HELP WITH IDENTIFICATION:

* *Basal sheaths = many short overlapping sheaths around bottom of shoot, at base of plant.*

** *Male spikes are at the extreme tip of the flowering stem (see drawing on page 123).*

- **Carnation Sedge** also has a **tough, thickened, 3-sided leaf tip**; the other species don't.
- **Stiff Sedge** also has a **very stiff flowering stem** - stiffer than those of the other species.
- **Carnation and Stiff Sedges** can look like, and share habitat with, **Green-ribbed Sedge** (page 128) but differ in their grey tinge; Stiff Sedge also differs from both Green-ribbed and Carnation Sedges in its pointed ligule (blunt in the other two) and shiny basal sheaths.
- **Carnation Sedge** can resemble, and grow with, **Tawny Sedge** (page 132) but Tawny Sedge is brighter yellow-green, with no grey tinge, and has 1-2 male spikes.

HABITAT/MANAGEMENT NOTES FOR SMALL TO MEDIUM-SIZED SEDGES WITH GREY-TINGED LEAVES

	Habitats	Management
Common Sedge *Carex nigra*	Mainly on **wet acid to neutral soils** in **mires** (esp. M1-8, 15-18, 23-26, 32 & 38), **dune slacks** (SD14-17), **grasslands** (U4-6) and some **woods** (esp. W4). Can form dense swards, especially in M6 mires.	**Rather unpalatable** to sheep, deer, cattle etc.
Glaucous Sedge *Carex flacca*	Mainly on **dry to damp basic soils** in **unimproved grasslands** (esp. CG1-6, 8, 9 & 13), **mires** (esp. M10, 11 & 13), **dune slacks** (SD13-17) and some **woods** (esp. W7). Grows mainly as scattered plants among other species; rarely forms dense, dominant swards or patches.	**Palatable** to grazing animal such as sheep, deer and cattle.
Carnation Sedge *Carex panicea*	Mainly on **damp, acid to basic soils** (esp. where **flushed**) in **unimproved grasslands** (esp. U4-5, CG9-13 & MC10), **heaths** (esp. H7 & M15/16) and **mires** (esp. M6-14, 22-26 & 37-38); commonest in **short** vegetation. Grows mainly as scattered plants among other species; rarely forms dense, dominant swards or patches.	**Palatable** to grazing animals (sheep, deer, cattle etc). **Tolerant of trampling**.
Stiff Sedge *Carex bigelowii*	Mainly on **damp to wet acid soils** on **mountains**, in **moss/rush summit heaths** (U9-10), **snowbeds** (U8/11-14/18), **dwarf shrub heaths** (H13/14/17/19/20/22), **grasslands** (U7/13), **bogs** (M19) and **flushes** (M7). Mainly in **short** vegetation. Can form dense swards.	**Moderately palatable** to sheep and deer. **Tolerant of trampling**.
Bottle Sedge *Carex rostrata*	Mainly on **very wet peaty acid to neutral soils** (M4-5 & M8-9 mires & S27 fen), in **woods** (W3) & **shallow open water** (S9 swamp). Can be dominant in dense or open swards on wet ground or in shallow water.	**Little grazed** (much habitat is too wet for access anyway)

Bottle Sedge with fruiting female spikes (June)

Bottle Sedge swamp

MORE PHOTOS OF SMALL TO MEDIUM-SIZED SEDGES WITH GREY-TINGED LEAVES

Stiff Sedge in short montane moss/grass/sedge vegetation in Aberdeenshire (June)

By late autumn Common Sedge forms swards of brown, narrow, rather curled leaves 20–40 cm tall, mixed here with the wider leaves of Glaucous Sedge, that have so far retained much of their grey-green colour but are starting to turn yellower. Photographed in East Lothian in November.

Green-ribbed Sedge *Carex binervis*

Quite distinct with its **rather broad (2-6 mm) dark green leaves** whose **tips turn red-brown** in autumn. **Short, blunt ligule**. Brownish female spikes quite **well-spaced** along stems 15-100 cm tall. Common on **dry to moist acid soils** in **heaths** (H8-10, 12 & 13-22) and **unimproved grasslands** (U4-7). **Rather unpalatable**, and **tolerant of trampling** by sheep, deer and humans.

The most similar species in these places are Carnation Sedge (page 124: paler **blue-grey** leaves), Stiff Sedge (mountain plant with **dull grey-green** leaves, **pointed** ligules and **stiff** stems), Common Cottongrass (page 137: **white cottony fruit heads** in summer, and leaves ending in long **3-sided points**) and Pill Sedge (page 134: smaller, with **narrow, yellow-green leaves** up to 2 mm wide and short flower spikes **clustered** at tip of thin stem 10-25 cm tall). Common and Glaucous Sedges (page 124) have **narrower** leaves (to 3 mm wide) that are **grey-green** on one or both sides. Tawny Sedge (page 132) can look rather like Green-ribbed Sedge but is smaller, brighter/yellower green, has a leaf tip abruptly tapering to a very fine extreme point, and grows in wetter, more basic habitats.

Leaves and flowering stem in June

Wood Sedge *Carex sylvatica*

A medium-sized, rather densely-tufted sedge with bright green leaves 2-6 mm wide and short (2 mm) blunt ligules. The female spikes are green, narrow and on long stalks; they can droop. Green-ribbed Sedge (page 128) has darker, stiffer leaves and wider, browner female spikes on shorter stalks. Wood Sedge grows on moist, neutral to basic soils in woods (esp. ancient W7-9). (Green-ribbed Sedge prefers acid soils.)

Plant in June

Remote Sedge *Carex remota*

A medium-sized woodland sedge with dense tufts of narrow (<2 mm wide) mid green leaves. Ligule short and blunt. The greenish female spikes are small, short, unstalked and spaced far apart; next to each one is a long bract (like a leaf). Grows on moist to wet, neutral soils in woods (esp. ancient W7).

Close-up of flowering shoot in June (many shoots in each tuft)

Large sedges of wet places

Many large sedges grow in wet habitats, and in some such places they can form dense patches. The individual species can look rather alike, especially as flowers or fruits are not always visible (commonly absent or only in places too wet to get to), so the table on the next page gives identification information based on leaves and general growth form. Three species are described elsewhere in the sedge section of this book - Bottle Sedge (page 125) and Greater Tussock-sedge & False Fox-sedge (page 130) - but are also included here too, for comparison with other species. The habitats of these large sedges are described below.

A typical large wetland sedge; this one is Slender Tufted Sedge

Slender Sedge *Carex lasiocarpa*	**Acid to basic, peaty swamps & mires**, esp. in **N & W** (esp. in NVC M1, M4-6, M9, M15, M17, S9, S27).
Fibrous Tussock Sedge *Carex appropinquata*	**Neutral to basic fens**. Scarce. Mainly in E Anglia & central Ireland. rare in N England & S Scotland.
Water Sedge *Carex aquatilis*	At **edges of rivers/lakes** & in **swamps/mires**, in **N & W**.
Tufted Sedge *Carex elata*	**Neutral to basic mires** (S1), **wet woods & watersides**. Rare in Scotland.
Bottle Sedge *Carex rostrata*	On **peaty acid to neutral soils** (M4-5/8-9 mires & S27 fen), in woods (W3) & **shallow open water** (S9 swamp).
Greater Tussock-sedge *Carex paniculata*	On **wet neutral to basic soils** in **woods** (W5) and **swamps** (S3) at **low altitudes**
Bladder Sedge *Carex vesicaria*	On **neutral to basic soils**, in **swamps** (S11), **wet woods** (W3) & **by rivers, streams and lakes**.
Slender Tufted Sedge *Carex acuta*	**Neutral to basic soils** in **swamps, fens, wet woods** and **watersides**, in the **lowlands**.
False Fox-sedge *Carex otrubae*	Mainly on **damp neutral soils** in **lowlands** and along **coastlines**; dense patches = NVC S18.
Lesser Pond-sedge *Carex acutiformis*	**Neutral to basic soils** in **swamps/fens** (S7), **wet woods** (W3/5/7) and **watersides**; mainly in the **lowlands**.
Cyperus Sedge *Carex pseudocyperus*	**Swamps/fens** (S17), **wet woods** and **watersides**, in **S Britain** and in Ireland.
Greater Pond-sedge *Carex riparia*	**Swamps/fens** (S6) and **neutral to basic waterside habitats** in the **lowlands**.
Pendulous Sedge *Carex pendula*	**Damp neutral to basic soils**, in **woods** (esp. W7) and along **ditches and streamsides**, in the **lowlands**.

See next page for identification notes >>>

IDENTIFICATION OF LARGE WETLAND SEDGES using leaf and growth form features

<<< See previous page for habitat notes	Leaf width	Leaf colour	Ligule length	Ligule tip	Growth form
Slender Sedge *Carex lasiocarpa*	1-2 mm	grey-green	2-3 mm	blunt	slightly tufted
Fibrous Tussock Sedge *C. appropinquata*	1-2 mm	yellow-green	2-3 mm	blunt	large tussocks
Water Sedge *Carex aquatilis*	3-5 mm	grey-green upperside; shiny yellow-green or dark-green underside	5-10 mm	blunt	tufted
Tufted Sedge *Carex elata*	3-6 mm	grey-green	5-10 mm	pointed	densely tufted
Bottle Sedge *Carex rostrata*	2-7 mm	grey-green upperside; dark green underside	2-3 mm	blunt	not tufted
Greater Tussock-sedge *C. paniculata*	5-7 mm	dark green	2-5 mm	blunt	v. large tussocks
Bladder Sedge *Carex vesicaria*	4-8 mm	yellow-green to mid-green	5-8 mm	pointed	not tufted
Slender Tufted Sedge *Carex acuta*	3-10 mm	grey-green	4-6 mm	blunt	tufted
False Fox-sedge *Carex otrubae*	4-10 mm	bright green	5-10 mm	pointed	densely tufted
Lesser Pond-sedge *Carex acutiformis*	7-10 mm	young = grey-green; older = dull green	5-15 mm	pointed	tufted
Cyperus Sedge *C. pseudocyperus*	5-12 mm	yellow-green	10-15 mm	blunt	slightly tufted
Greater Pond-sedge *Carex riparia*	6-15 mm	grey-green	5-10 mm	blunt	tufted
Pendulous Sedge *Carex pendula*	15-20 mm	bright green upper-side; greyer underside	30-60 mm	pointed	tufted

Pendulous Sedge in May, with distinctive large, drooping female flower spikes (a feature also seen in Cyperus Sedge)

Slender Sedge in July: note the narrow long, leaves with curled tips

Yellow-sedges *Carex demissa/lepidocarpa/viridula*

Small sedges with tufts of **bright green leaves 2-5 mm wide**, and **short, spiky-looking yellow-green female spikes**, each one on a very **short stalk. Leaf-like bracts** grow on main stem (one for each female spike). On **wet acid to basic soil**, in **flushes & other mires** (esp. M9-12, 34 & 37-38). Also on stony **loch shores**. Mainly in vegetation kept **short**, **grazed or trampled** vegetation.

Common Y-S *C. demissa*, the commonest species, has leaves up to 20 cm long x 5 mm wide; lowest female spike commonly much further down stem from other flower spikes; fruit with projecting beak c. 1 mm long; in acid to basic mires. **Long-stalked Y-S** *C. lepidocarpa* has leaves up to 8 cm long x 5 mm wide; all female spikes close together near top of stem; fruit with projecting beak 1.5-2 mm long, bent backwards; in mires on basic soils. **Small-fruited Y-S** *C. viridula* has leaves up to 8 cm long x 3 mm wide; female spikes all near stem tip; fruit with projecting beak <1 mm long; uncommon; mainly loch shores, coasts and dune slacks.

Close-up of female spike, showing fruits with projecting 'beaks'

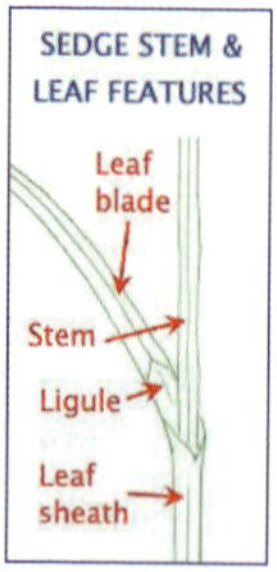

Tawny Sedge *C. hostiana* has similar **bright green leaves** 2-4 mm wide but **tapered quite abruptly to a final very narrow extreme tip 5-15 mm long**. The female spikes are well spaced along stems that grow 20-30 cm tall; the plant can look like Carnation Sedge (page 124) but yellow-green instead of grey-green. Grows on **wet basic soils**, in **flushes & other mires** (esp. M9-12). Mainly in **upland areas**, in short, **grazed** vegetation.

IDENTIFICATION USING BASAL LEAVES ALONE. Yellow & Tawny Sedges: <8 veins each side of central vein; Tawny Sedge leaf tapers abruptly to very narrow extreme tip. **Broad-leaved Cottongrass** (page 137): looks similar but >8 veins each side of central vein; stout 3-angled tip.

Star Sedge *Carex echinata*

A medium-sized sedge with **dense tufts** of **mid green, narrow leaves 1-2.5 mm wide**. Leaves look nondescript but are quite **thick** (relative to their width) and have a tip that is **3-angled** in section. Easily told in summer by the 3-4 short, unstalked, **star-like spikes** with spreading, pointed, yellow-green to yellow-brown fruits; these spikes are closely spaced along the final 1-3 cm of the stem tip. Stems 20-40 cm long, straightish, with no leaves or obvious leaf-like bracts in their upper parts. Yellow Sedge (page 132) has similarly spiky, yellow-green spikes but has wider leaves, and distinct leaf-like bracts well up the stems. Star Sedge grows on **wet acid soils**, especially in **flushes** (M6-8, M12 & M15), **bogs** (M17-21), **Purple Moor-grass mires** (M25-26) and **wet woods** (W4). It is **not very palatable** to sheep, cattle and deer. Habitats vary from lightly to rather heavily grazed.

False Fox Sedge *Carex otrubae* has a flower head that is similar but **larger (8-14 cm long) with more flower spikes** on a **stem that is very thick and strongly 3-angled**. The **bright green leaves are much wider** (4-10 mm) and the **whole plant is large and tufted**. Mainly on **damp neutral soils** in **lowlands** and along **coastlines**.

Greater Tussock-sedge *Carex paniculata* has a **large flower head 5-15 cm long**, again with more flower spikes. It forms **unmistakeably huge tussocks** up to 1.5 m tall, with dense masses of **very long, dark green leaves** 4-7 mm wide. It grows on **wet neutral to basic soils** in **little-grazed woods** (W5) & **swamps** (S3) at **low altitude**.

Star Sedge flower head in July

False Fox Sedge in July

Greater Tussock-sedge in May

Pill Sedge *Carex pilulifera*
Spring Sedge *Carex caryophyllea*

Small, tufted sedges with **low rosettes of narrow leaves** up to 2 mm wide, and short (5–8 mm) green-brown **flower spikes clustered at tips of thin stems 10–25 cm tall**: male spike at extreme tip; female spikes just below. Mainly in short, **grazed** vegetation on **dry soils**. **Palatable** to sheep, deer and cattle.

Pill Sedge *Carex pilulifera* has **yellow-green leaves**, and the **flower stem commonly curves outwards and droops at the tip** because of the weight of the flower head at the end of such a thin, relatively weak stem. The **male spike (at extreme tip of stem) is thin and pointed**. Mainly on **dry, acid soils**, in short, **grazed, unimproved acid grasslands** (U4–7; also some CG11, which is more basic) and **heaths** (esp. H10, H12–14 & H16–20). Also in **mountain moss heaths** (U10) and **woodland clearings**. Common on **disturbed** ground: for example after **burning** or **tree-felling**.

Spring Sedge *Carex caryophyllea* has **darker leaves**, **reddish sheaths at the plant base** (brown in Pill Sedge), a **straighter and more upright flower stem** and a **wider male spike with a rather blunt tip**. On **dry, neutral to basic soils**, mainly in short, **moderately to heavily grazed, unimproved grassland** (CG1–2 & 7–10, SD8 & some U4). Short (1–2 mm) ligule helps separate sand dune specimens from Sand Sedge (page 135).

Pill Sedge in May

Spring Sedge in May

Pill Sedge in October

Pill Sedge in May

Sand Sedge *Carex arenaria*

A **medium-sized** sedge with many **tufts of mid green leaves** up to 60 cm long and 4 mm wide, bluntish ligules up to 5 mm long, and at the top of the stem a **rather dense, golden-brown flower head** up to 4 cm long. The flowering stems are up to 60 cm long, but many tufts have leaves only. Grows in **loose sand in coastal dunes**, commonly with Marram (page 120). Spreads by means of **underground shoots** (rhizomes) from which arise many individual leafy tufts that can grow in **straight lines** (see photo below). Short, non-flowering plants within denser swards (SD8) can look like Spring Sedge (page 134) but that species has leaves only up to 20 cm long, and ligules just 1–2 mm long.

Leaves

Flower head, with Marram leaves behind

Line of shoots arising from an underground stem (rhizome). Paler leaves are Marram.

Brown Sedge *Carex disticha* looks similar but has **denser, darker brown flower heads, wider leaves (up to 6 mm) and longer ligules (up to 8 mm)**, and grows on **damp to wet, more or less neutral soils** in **mires** (esp. M22–23 & 27) and **dune slacks** (esp. SD17).

Oval Sedge *Carex leporina*

A tufted plant rather like Star Sedge but with leaves wider (2–3 mm) and distinctive **brown, dense, oval flower spikes clustered at the stem tip**. Ligule only about 1 mm long. On **moist acid to neutral soils**, in **grasslands** (e.g. U4 & MG5) & **heaths**, especially where the vegetation is not too tall and dense (e.g. where **trampled by cattle or sheep**); also on **barer disturbed ground and along tracks**.

White Sedge *Carex canescens*

Another species resembling Star Sedge in its tufted habit, but again with leaves wider (2–3 mm). Similar to Oval Sedge in this respect, but ligule longer (2–3 mm) and **flower head with short, neat, oval, pale green spikes**. On **wet acid soils** in **mires** (esp. M6–7 sedge mires); mainly in the N and W.

White Beak-sedge *Rhynchospora alba*

A rather small sedge forming narrow tufts with very narrow leaves. Flower head small but distinctive in its **spreading and flattened shape**, and its colour: **whitish in summer, becoming pale brown in autumn** (and persisting quite well through winter). Grows on **wet, acid peat** in **bogs** (M1/2, M17–18 & M21) and **wet heathy flushes** (M15a). Mainly in the west.

Brown Beak-sedge *R. fusca* is similar in appearance and habitat, but has a **dark red-brown flower head** and the thin leaf-like bract sticking up/out at the stem tip is **at least twice as long** as the length of the adjacent flower head (in White Beak-sedge this bract is either shorter or just a little longer than the flower head). A cluster of sites in the New Forest area but otherwise very thinly scattered in W Scotland, Ireland and W Wales.

Common Cottongrass *Eriophorum angustifolium*

The **white cottony fruit heads**, conspicuous in summer, give Cottongrasses their name. Common Cottongrass has shiny **dark green leaves 3-6 mm wide**, tapering to **long (5-10 cm), thickened, 3-sided, pointed tips**. The leaves are reddish-tipped in autumn and eventually turn **reddish-brown** throughout. There are pink leaf sheaths at the base of the plant. The **white fruit heads are in drooping clusters of 3-7 at the stem tips** and are most abundant where the vegetation is **not heavily grazed**. Grows in **wet** peaty habitats: acid **bogs** (M1-3 & M17-21; especially abundant in M3 **bog pools** and **heavily trampled or wetter cut-over areas**), **wet heaths** (M15-16) and various **sedge mires** (esp. M4-9). Unlike Hare's-tail Cottongrass it is not restricted to bogs. It is **unpalatable** to grazing animals such as sheep, cattle and deer.

This photo and the upper photo: plants in July

Plants in a bog in October

Broad-leaved Cottongrass *E. latifolium* has **shorter yellow-green leaves** with **3-sided tips <5 cm long**, and **rough-stalked** fruiting heads (smooth-stalked in Common Cottongrass); a rather uncommon plant of **basic flushes** (esp. M9-11). Its basal leaves resemble those of Yellow & Tawny Sedges (page 132) but with thicker 3-sided tips and >8 veins each side of the central vein.

Hare's-tail Cottongrass *Eriophorum vaginatum*

This species forms **dense tufts of thin, wiry leaves** up to 50 cm long, arising from the base of the plant. Each flowering stem has shorter leaves whose wiry blades project from very inflated sheaths, and **one upright** oval spike of tiny yellowish flowers in spring, becoming a conspicuous **white cottony fruit head in summer**. Flowers and fruits are most abundant where **not heavily grazed**. Grows mainly on **deep, wet acidic peat** in **bogs** (M17–20): a good indicator of deep peat bog habitat. **Unpalatable** to grazing animals such as sheep, deer and cattle. Can become dominant (M20) where **overgrazing** has eliminated dwarf shrubs. Differs from Deergrass (page 140) in having tufts of true leaves (not stems) that are dull brown (not golden) in autumn–winter, and in its white cottony fruit-heads in summer. Differs from wiry-leaved grasses in the white cottony fruit-heads, 3-angled leaf section and the very inflated sheaths of the stem leaves. Black Bog-rush (page 141) differs in its very dark flower heads and the blackish, shiny leaf sheaths at shoot bases.

Flower head in April

Plant in September (tuft of long dull green leaves)

Plants in late May, with white cottony fruiting heads

Sedges with a single small, narrow flower spike

Plants with **upright stems**, each with a **small, narrow oval, brownish flower spike at its tip**. Deergrass is included here but also described more fully on page 140.

IDENTIFICATION (see also drawings below, showing leaf sheaths and flower spikes)	Growth form	Width of stem	Length of flower spike	Lowest scale/unit of flower spike, as % of length of whole spike	Shape of top of leaf sheath	Length of leaf blade
Common Spike-rush *Eleocharis palustris*	Stems grow singly or in small tufts	1-4 mm	5-20 mm	<20%	straight	No leaf blade
Slender Spike-rush *Eleocharis uniglumis*		<1 mm	5-12 mm			
Few-flowered Spike-rush *E. quinqueflora*		<1 mm	5-7 mm	>50%	angled	
Many-stemmed Spike-rush *Eleocharis multicaulis*	Dense tufts, each with many stems	1-1.5 mm	10 mm	c.30 %		
Deergrass *Trichophorum germanicum*		1-1.5 mm	3-6 mm	100%		0.3-1 cm
Dioecious Sedge *Carex dioica*	Small tufts	<1 mm	5-20 mm	<30%		5-20 cm
Flea Sedge *C. pulicaris*	Dense tufts	<1 mm	10-25 mm	<25%		5-25 cm

MAIN HABITATS	
Common SR *Eleocharis palustris*	**Neutral** mires, fens, swamps (esp. S19), ponds etc
Slender SR *E. uniglumis*	Wet **basic** soils, especially along **coasts** (e.g. SM20)
Few-flowered SR *E. quinqueflora*	Short, **grazed**, **base-rich flushes** (esp. M10-12)
Many-stemmed SR *E. multicaulis*	**Acid** mires, wet heaths & bogs (M6-7, 15-18 & 21)
Deergrass *Trichophorum germanicum*	**Acid peat**, esp. in wet heaths & bogs (M15-18)
Dioecious Sedge *Carex dioica*	Short, **grazed**, **base-rich flushes** (esp. M10-11)
Flea Sedge *C. pulicaris*	**Base-rich flushes** (M10-13) & **grazed damp grassland** (CG9-11)

Deergrass *Trichophorum germanicum*

Forms **dense tufts of wiry, rather stiff 'leaves'**, as in Hare's-tail Cottongrass, except that the 'leaves' are actually **stems** c.1 mm wide and up to 40 cm long, each one topped with a **small brownish flower spike** in spring and summer. When flowers/fruits have gone the **stem tip is left blunt and rough**, unlike the clean, finely-pointed leaf tip of Hare's-tail Cottongrass. The leaves are tiny (<1 cm long), on the lower parts of the stems. Turns **orange-brown** in autumn, when Hare's-tail Cottongrass is duller and greyer. Grows on **moist or wet, acidic peat** in **bogs** (mainly M17-18) and **wet heaths** (M15-16). **Unpalatable** to grazing animals such as sheep, deer and cattle. Can be very abundant where dwarf shrubs have been reduced or eliminated by **grazing** or **burning**. **Tolerant of trampling** by grazing animals and by people.

Plants in September, with shoots turning orange-gold at tips

Deergrass flowers in May

Shoot tips (+ old fruit heads) in September

Northern Deergrass *T. cespitosum* is similar but the **tufts are less dense**, the **stems are thinner** (<0.7 mm) and **can be slightly curved**, and the **angle of the top of the leaf sheath is less steep** (<c.50°). In **wet bogs** (M1/2/17) and **base-rich flushes** (M10/11) in N Britain and in Ireland. Sterile hybrids are common. For more identification details see https://www.edencroft2.co.uk/deergrasses.html

Black Bog-rush *Schoenus nigricans*

This species forms **dense tufts** of **shiny green wiry stems (up to 60 cm tall)** and **thin wiry leaves**, all becoming very **pale in winter**. **Blackish, shiny leaf sheaths** at bases of shoots. **Blackish-brown flower head at stem tip**. A long leaf-like bract protrudes from the base of the flower head, sticking out at an angle from the main stem. Commonest in the west. Mainly in **flushes, fens & flushed wet heaths** (M10-11, M13-15, M25 & H5-6) on **wet, more or less basic soils**, but in extreme W Scotland & Ireland also in acid **bogs** (M17), **flushed coastal grassland** (MC10) and on damp **sea cliffs** (vegetation related to M25). Occurrences in western bogs might indicate enrichment by wind-blown chemicals of marine origin. Appears to be **quite palatable** to animals such as sheep, cattle and deer. **Heavily grazed** plants typically short, with many shoots and leaves bitten back. **Lightly grazed** plants can form tall, dense tussocks.

Black Bog-rush in wet heath (NVC M15) in August

Saltmarsh Flat-Sedge *Blysmus rufus* has a **dark, dense flower head** but is **short** (mostly <30 cm), is **not tussocky**, and forms **patches** (SM19) on **saltmarshes** around the coasts of N Britain and N Ireland. The related **Flat-sedge** *B. compressus* looks similar but has **flat leaves up to 4 mm wide** (inrolled in *B. rufus*) and grows in **neutral to basic wetlands** in England (rare in Wales and S Scotland).

Flat-sedge in July

Sea Club-rush *Bolboschoenus maritimus* ▸ also has conspicuous **dark, dense flower heads** but is **tall** (up to 1 m), **patch-forming** (not tussocky), and has **long leaves up to 7 mm wide**; leaves are at plant base and on the stems. Leafy bracts overtop the flower head. Grows on **wet muddy ground** and in **shallow water** in **brackish habitats** around most parts of the British and Irish **coasts**. Can form **dense stands** (S21).

Sea Club-rush in September

Section 2c

RUSHES

This is a varied group of hairless, grass-like plants with small, brown or green flowers in dense or open clusters. Our commonest and most conspicuous rushes are distinctive enough not to be confusable with other plants.

Four species - Soft Rush, Compact Rush, Hard Rush and Sea Rush - have dense tufts of tall stems, each flowering stem with a brown flower head sticking out sideways some distance down from the pointed stem tip. The leaves of Soft, Compact and Hard Rushes are very small, inconspicuous, dull brown sheaths on the lower parts of the stems. (Sea Rush has longer 'proper' leaves.)

The other five species described here have branched flower heads at the stem tips and long leaves from the plant base and (except in Heath Rush) the stems. Sharp-flowered, Jointed and Blunt-flowered Rushes have thick, hollow leaves with short internal cross walls at intervals (run a finger & thumb along a leaf and the cross-walls feel like little bumps). Saltmarsh Rush is smaller with fine, slender stems, and grows in saltmarshes. Heath Rush has very stiff, leafless stems and unmistakeable low, dense rosettes of very stiff, glossy green leaves.

Collectively these rush species occupy a wide range of habitats but are most common on damp or wet soils in mires, flushes, grasslands, woods and saltmarshes, though they are comparatively scarce in bogs.

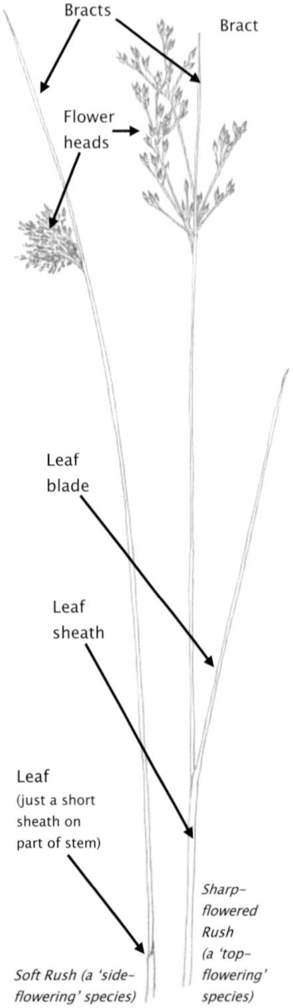

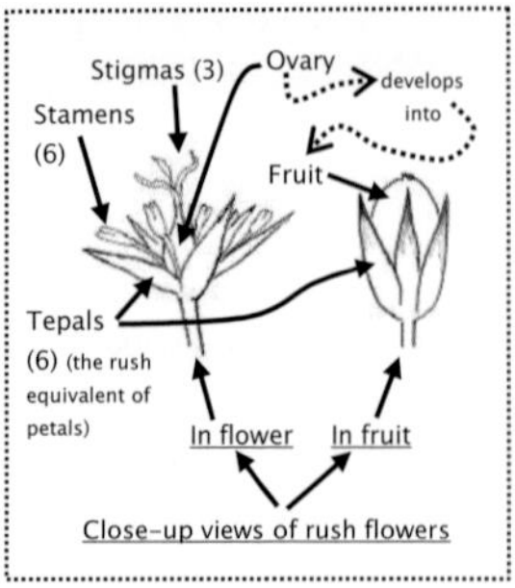

'Sideways-flowering' rushes

Tall, densely tufted plants up to 1m or more in height, with brown flower heads sticking out from the side of the stem some distance down from the pointed stem tip. The leaves are very small, inconspicuous, dull brown sheaths on the lower parts of the stems (except in Sea Rush whose leaves are long, like those of the species on the next page).

Soft Rush (October)

Soft Rush *Juncus effusus* has smooth green stems (vertical grooves present but so shallow that stem looks and feels smooth), and fairly dense mid brown flower heads. Variety *compacta* has denser flower heads as in Compact Rush but paler brown. Very common on damp or wet acid to neutral soils in mires/wet grasslands (esp. MG10, M6 & M23), heavily grazed or poached grassland (U4 & MG5-6), wet woods (W1, W4 & W7) and disturbed places such as felled woodland and tracksides.

Compact Rush *J. conglomeratus* has slightly greyer and distinctly (vertically) grooved stems (grooves easy to feel), and very dense, dark brown flower heads. In similar habitats to those of Soft Rush, but generally not as common as that species.

Hard Rush *J. inflexus* has thin but stiff, vertically grooved, grey-green stems and more open, spreading flower heads. In damp grasslands/mires (esp. MG10b) on moist, neutral to basic soils in lowlands. Widespread, but uncommon N of Dundee.

Compact Rush (dense flower head)

Soft Rush (intermediate flower head)

Hard Rush (more open flower head)

Sea Rush *J. maritimus*: like Hard Rush (open flower head) but greener (less grey) and with long, sharply-pointed leaves. In saltmarshes (esp. SM15+18) but rare from NW Scotland to Yorkshire.

'Top-flowering' rushes

These and Heath Rush (page 145) have their small brown flowers in **branched flower heads at the tips of the stems**. The species on this page have **long leaves arising from the plant base and up the stems**, and the leaves of the first three species on this page **are thick (up to 3 mm wide) and hollow, with short internal 'cross walls' at intervals** (run finger & thumb along leaf and the 'cross-walls' feel like little 'bumps').

Above: Sharp-flowered Rush sward in July.
Left: leaf cut open to show internal cross-walls

Sharp-flowered Rush *Juncus acutiflorus* is patch-forming, with **tall (30-80 cm), straight** stems with **straightish cylindrical leaves up to 3 mm wide**. The flower head is **large and well-branched**: most branches point upwards. The small brown fruits are **sharply pointed**. It grows on **deep, wet, acid to neutral soils** in **mires** (esp. M6d & M23a) and **woods** (esp. W4 & W7). Can hybridise with the next species.

Jointed Rush *J. articulatus* is **shorter (15-40 cm)** and **tufted**, with **curved stems** (growing out at base, then curving upwards) and **curved leaves** that are **flattened in cross section**. The fruits are **blunt with tiny, very short points**. Grows on **moist, neutral to basic soils** in **mires, flushes** (including stony, neutral to basic M10-11 flushes where Sharp-flowered Rush is absent), **damp grassland** & **disturbed ground.**

Blunt-flowered Rush *J. subnodulosus* is like Sharp-flowered Rush but is **paler, brighter green**, the **leaf blades have internal longitudinal walls as well as transverse cross-walls**, the flower head has **wider-angled branching** (mixture of upwards-, sideways- & downward-pointing branches), and the fruits are **blunt**. On **wet, neutral to basic soils in southern British lowlands**, especially in **rushy mires** (M22).

Saltmarsh Rush *J. gerardii* is **short, slender** and patch-forming (not tufted), with **thin wiry stems 10-30 cm tall, and narrow leaves <1.5 mm wide** (arising at base of plant and on stems) that **lack internal cross-walls**. Common on **saltmarshes.** Sea Rush (previous page) also grows in saltmarshes but is much bigger (to 1 m tall), and stiffer, with sharply-pointed stem and leaf tips and 'sideways' flower heads.

Heath Rush *Juncus squarrosus*

Unmistakeable. **Dense, low-grown** (even 'sunken') **rosettes of very stiff, glossy, dark green leaves** up to 15 cm long, 2 mm wide and with a groove running up the middle of the upperside. **Very stiff flower stems** 15-30 cm tall, **topped with branched brown flower heads**. Grows on **moist acid soils**, mainly at **moderate to high altitudes**. Forms dense swards in **rush-heaths** (U6); also in **grasslands** (U4-5 & U7), **dwarf shrub heaths** (M15-16), **bogs** (M17-20) and some herb-rich neutral flushes. **Unpalatable** to grazing animals such as sheep, deer and cattle. Mainly in short, **heavily grazed vegetation** where more palatable species are preferentially grazed, giving Heath Rush a stronger competitive edge. 'Clean' rosettes (see photo below) suggest heavier grazing than where partly grown through with other plants.

Basal rosette of leaves (June)

Flower heads (June)

Brownish *Juncus squarrosus* rush heath (U6) and paler *Nardus* grassland (U5) in August

Three-leaved Rush *Juncus trifidus* ►

Short, dense **tussocks** of **thin wiry green to gold-brown stems, leaves**, and **bracts overtopping very small brown flowers**. On **thin/stony acid soil** on **high montane summits and plateaux** in the Scottish Highlands and islands. Locally abundant in rush heath (U9) - e.g. in the Cairngorms.

Two small rushes forming branched tufts with short wiry leaves and small flowers.

Bulbous Rush *Juncus bulbosus.* Stems and branches widely spreading; overground stolons can take root at the nodes; plant can have a bulbous base; stems can be reddish-brown; flowers/fruits mid to dark brown and can sprout young leaves/plantlets. On wet acid/neutral soils in heaths, bogs, shallow acid pools (A24), periodically flooded places, by tracks, etc.

Toad Rush *J. bufonius.* An annual. Branches less spreading than in Bulbous Rush. Flowers/fruits mainly greenish, with finely-pointed tepals. On periodically damp/wet soils and on disturbed ground (e.g. by roads and tracks).

Section 2d

WOODRUSHES

Woodrushes have flat leaves 2-12 mm wide, like those of grasses, growing at the base of the plant and up the stems, and dense to open clusters of small brown, rush-like flowers.

A distinctive feature not seen in grasses, sedges or rushes is the occurrence of long white hairs along the edges of the leaves.

Despite the name of Woodrush, the group as a whole occupies a wide range of habitats: woods, grasslands, heaths, bogs and mountains.

The identification of our five commonest woodrush species is generally straightforward:

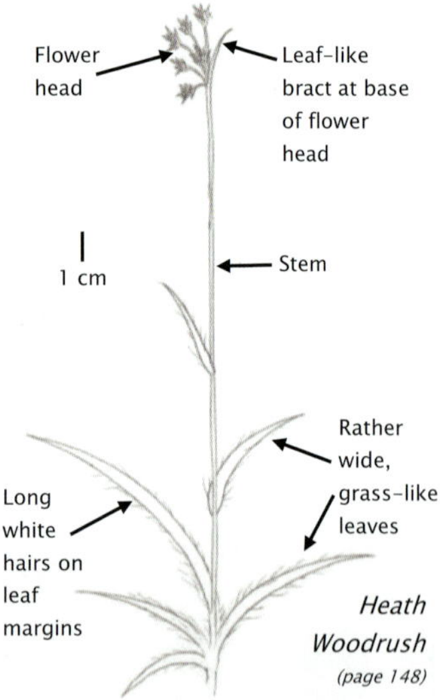

Greater Woodrush is large, with long, broad (6-15 mm wide), shiny, bright green, tough-textured leaves. It can form extensive carpets. **Hairy Woodrush** is like a small version of Greater Woodrush but with softer leaves 5-10 mm wide, tapering to their bases as well as their tips. **Southern Woodrush** is like Hairy Woodrush but with narrower leaves and drooping flower heads. **Many-headed Woodrush** and **Field Woodrush** are small, with duller green leaves that do not taper to the base; separated by details of leaf width, stem length, and flowers. See next two pages for more details about all five species.

Greater Woodrush *Luzula sylvatica*

A **robust, tufted** plant with **broad, pointed, tough, shiny, bright green leaves 6-15 mm wide and up to 30 cm long**, **firmer-textured and more shiny** than in most grasses, and with **long white marginal hairs**. They can be evergreen. Dead brown leaves persist for a long time. The **large, branched flower head** is at the tip of a stem 30-80 cm long, and has many small, brown flowers.Grows mostly on **dryish to moist, acid soils** in **woods** (esp. W10-11 & 15-17) and on **hillslopes and cliffs** (U16-17). Also in grasslands (esp. U4-6), heaths (esp. H12, 18 & 21-22) and bogs (esp. M19). In southern lowlands mostly in **ancient** woods. **Palatable** to grazing animals such as sheep and deer. Can form extensive, lush, almost pure swards where only **lightly grazed**; more grazing generally results in more grass and less woodrush.

Flower head in May

Leaves (partly grazed) in May

Flowering sward of Greater Woodrush in an oakwood in May

Smaller Woodrushes *Luzula* species

These show the typical Woodrush feature of long white hairs along the leaf margins. They also have a tiny 'knob' at the extreme leaf tip.

Hairy Woodrush *Luzula pilosa* has **bright green, sparsely hairy lower leaves** 5-10 mm wide and up to 15 cm long, **tapering noticeably to the base as well as the tip**. **Flower head very open** with branches subdivided and sticking out at a wide angle as in Greater Woodrush (page 147) on upright stems 15-30 cm tall. On **dry to moist acid soils** in **woods** (esp. W11 & W17-19) and **hedgebanks**; also in some **heaths** (e.g. H12). In southern lowland areas grows mostly in **ancient** woods.

Hairy Woodrush (May)

Southern Woodrush *L. forsteri* is similar but **lower leaves <4 mm wide and flower head drooping to one side**. **Acidic woods and hedgerows in S England & SE Wales**.

Heath Woodrush *L. multiflora*. See drawing on page 146. This has **dull green, hairy, parallel-sided leaves** up to 6 mm wide and mostly <20 cm long. The **small flower heads** are on upright stems 10-30 cm tall and vary from compact to open with short **straight** undivided branches. Flowers April-June. On **dry to moist, acid soils**, in short, **grazed, unimproved grassland** (esp. U4-6) and woodland (esp. W11).

Field Woodrush *L. campestris* is like Heath Woodrush but with creeping above-ground shoots (stolons), **shorter** stems (to 15 cm), **narrower leaves** 2-4 mm wide, and flower heads with **curved** branches. Flowers between March and June (flowering starts earlier than in Heath Woodrush). Mainly on **dry to moist, basic to mildly acid soils**, in short, **grazed, unimproved grassland** (MG3/5, CG7/10/13 & U4).

Hairy Woodrush leaves (January)

Field Woodrush (late March)

Section 3

OTHER PLANTS

These are split into six groups, based mainly on features of their leaves:

Leaf divisions vary from simple side lobes (e.g. dandelion) to intricate, deeply cut patterns.

Leaves with 3 or more lobes or leaflets radiating out from a central point.

A whorl is a group of 3 or more leaves, all arising at the same point along the stem.

An easy group to recognize.

Note: species with leaves only at the base of the plant (i.e. not on the stems) also belong in this group.

These are mosses, liverworts, filmy ferns and clubmosses: short plants with tiny leaves (or just a simple flat, tongue-like 'thallus') and no flowers. Page 304 describes the differences between mosses, liverworts, filmy ferns & clubmosses. A few lichens are described very briefly on p. 337, but lichens & fungi are not in the Plant Kingdom.

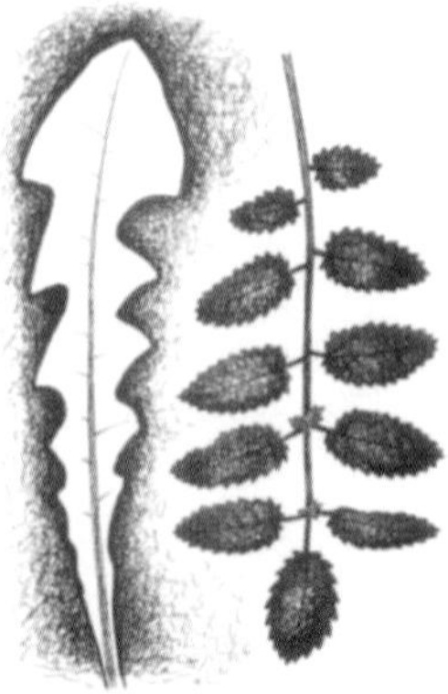

Section 3a

LEAVES WITH LOBES OR LEAFLETS IN ROWS ALONG LEFT & RIGHT SIDES (PINNATELY DIVIDED)

These are herbs or ferns whose leaves are either cut into **lobes** along the left and right sides (as for example in the left hand leaf in the drawing above) or divided more deeply into separate **leaflets** along a central leaf stalk (as in the right-hand example above). Some species have leaves divided into leaflets on a finer scale, as in the drawing below:

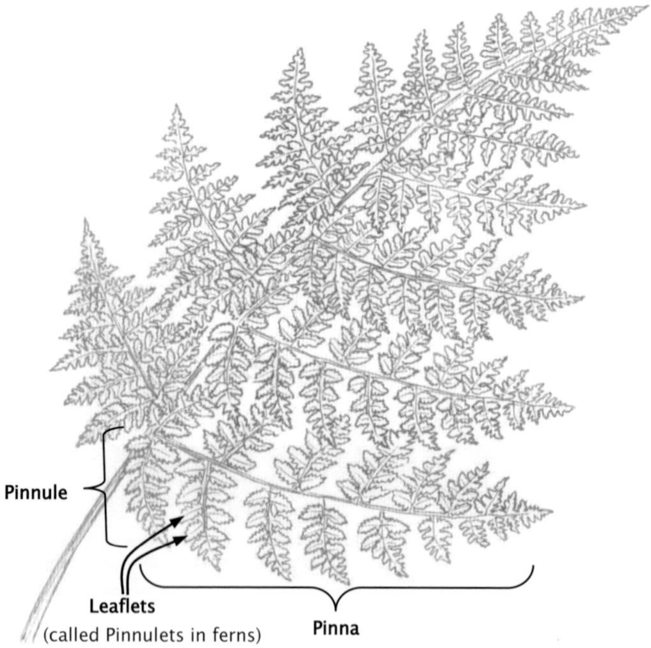

This section starts with ferns - all but three of the fern species described in this book (the other three are on pages 267 and 335). Ferns are primitive plants whose leaves are more correctly called fronds. Each frond is on its own stalk growing straight from the base of the plant. There are no other stems. Ferns lack flowers but produce spores in small brown patches, mainly on the undersides of fronds.

The rest of this section consists of herbs, starting with those whose leaves are divided into lobes (i.e. leaves with 'wobbly edges'), then those with leaves cut into very thin segments that are so thin that it isn't always clear whether they are best described as lobes or leaflets, and ending with those whose leaves are clearly divided into many separate leaflets.

Greater Knapweed (page 164) can deviate from the section's definition in having some undivided leaves. Oil-seed Rape and similar species (p. 268), Common Knapweed (p. 276), Melancholy Thistle (p. 277), Nipplewort (p. 284) and Oxeye Daisy (p. 286) are in section 3e (undivided leaves) but can have some leaves lobed.

Creeping Buttercup is described on page 200 in section 3b because it typically has three main lobes, but some leaves (see photo) would fit here in section 3a because they have five lobes: two opposite pairs + one end lobe.

How to tell ferns from herbs with intricately divided, fern-like leaves

The herbs in question are mainly umbellifers (pages 176-181) such as Cow Parsley. An example of the visual similarity between certain fern and umbellifer leaves is shown in the left and middle photographs below:

Spore cases on underside of fern

Fern stalk with brown scales

If the plant has flowers, or if any of its leaves grow out from a stem (both of these features visible in this photo of Cow Parsley), it is not a fern. If the leaves have small brown dots (spore cases) on their undersides, or if the leaf stalks have brownish scales, the plant is a fern. If the plant has only basal leaves, and these lack spore cases or scales, it could be in either group so look at the bottom of the leaf stalk. In umbellifers the base of the leaf stalk broadens into a pale sheath inside which are more (younger) leaves or a young developing stem. The base of a fern leaf stalk does not form a sheath.

Sheathing bases of umbellifer leaf stalks

Hard Fern *Blechnum spicant*

Easily told by its **tufts of long, narrow, stiff, glossy fronds** ('leaves') **with undivided, untoothed side lobes**. Bright green in spring; darker by late summer. Fertile fronds have very narrow side lobes. Grows on **moist acid soils** in cool, humid places, especially **woods** (W11 & 17) and N-facing **heaths** (esp. H10 & 21). **Palatable** to grazing animals: a good indicator of **grazing** (bitten off, square-ended frond tips distinct from pointed ungrazed ones). Polypody (page 153) looks similar but is not tufted, stiff or glossy, does not have separate fertile and sterile fronds, and grows on rocks, walls and trees.

Plant in April (green fronds are last year's growth)

Deer-grazed fronds (June)

Fertile, spore-bearing frond (April)

Polypody *Polypodium* species

Polypody is like Hard Fern in having **fronds divided only once**, but differs in being **not tufted** and **not glossy**. Fronds are up to 30 cm long. Grows mainly on **rocks, walls and trees**. Three species: Common and Western Polypody are widespread and common; Southern Polypody is south-western and rather uncommon.

Frond 3–7 cm wide; parallel-sided for most of its length; side lobes with blunt tips and untoothed edges

Lowest pinnae not bent back

Common Polypody
Polypodium vulgare

Frond widest (8–14 cm) ⅓ to ½-way up from base; side lobes with blunt or pointed tips and toothed or untoothed edges

Western Polypody
Polypodium interjectum

Frond widest (9–12 cm) near base; side lobes with blunt or pointed tips, and toothed edges

Lowest side lobes bent back

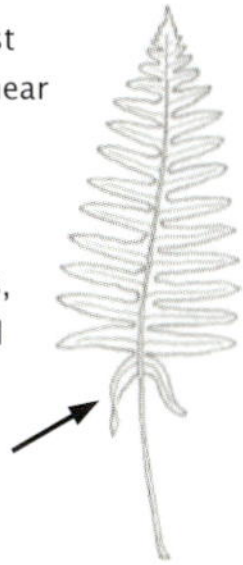

Southern Polypody
Polypodium cambricum

If small brown spore-cases are present on the undersides of fronds, check their shape: round in Common Polypody and oval in the other two species.

Maidenhair Spleenwort *Asplenium trichomanes*

A **small, tufted, evergreen fern**, each frond with a **blackish central stalk** and many **oblong, blunt-ended 'leaflets'** (pinnae) **with narrow bases**. Common on **steep, dry, acidic to basic rocks** and **walls** in **well-lit or shaded** places.

Maidenhair Spleenwort

Green Spleenwort *A. viride* is similar but with a **green central stalk**, so the whole plant is green. In **moist, shaded or sheltered crevices of basic rocks in upland areas**: **habitats include woods, cliffs and in limestone pavements**.

Rustyback *A. ceterach* also has a **green central stalk** but the **side lobes** (pinnae) **have broad bases and red-brown undersides**. On **basic rocks and mortared walls** in the **SW half of Britain and in Ireland**.

Rustyback

Three more small ferns of rocky places

Wall-rue *Asplenium ruta-muraria* ► is a **small, tufted fern** of **basic rocks** and **walls**. Its slightly **bluish-green** fronds are up to 12 cm long and divided and sub-divided further, with all **final (smallest) divisions diamond-shaped**.

◄ **Black Spleenwort** *Asplenium adiantum-nigrum* is another **small, tufted fern** of **basic rocks** and **walls**. Its **shiny green triangular, pointed fronds** are up to 25 cm long, **regularly divided** and **widest at the base**. The **lower parts of the stalks can be blackish**.

Brittle Bladder-fern *Cystopteris fragilis* ► is rather like Black Spleenwort but is **not tufted** and **not shiny**, has a **delicate texture** and has **veins branched and running out to the extreme leaf margins**. Side branches (pinnae) quite well spaced apart, especially toward base of frond. In **shaded crevices of basic rock** in **upland** areas.

Parsley Fern *Cryptogramma crispa*

An unmistakeable species whose **intricately divided fronds form dense tufts** up to 30 cm tall on **acid** soils, mainly among **screes** (U18 where snow lies late; U21 elsewhere) but also among rock outcrops. A plant of **uplands**, but rare in Ireland. The **fronds are very finely divided** and further sub-divided; the final (smallest) divisions are diamond-shaped and toothed on sterile fronds (which lack spore cases) or narrow oval and untoothed on fertile fronds (which have brown spore cases on undersides). Sterile fronds **look very much like Parsley.**

Plant growing in scree (Lowther Hills, August)

Closer view of mixture of fertile and sterile fronds

TALL TUFTED FERNS: INTRODUCTION

This page and the next two pages cover several **tall, tufted ferns** growing **mainly in shaded habitats**. This page has photos of some of the species. Identification details are given on the next page, followed by a page of habitat details. Narrow Buckler-fern is not actually tufted but is included here because of its other similarities with certain closely related tufted species.

Lemon-scented fern (September)

Male fern (May)

Soft Shield-fern (April)

Hard Shield-fern (last year's frond in April)

Scaly Male-fern (June)
One side of frond grazed by deer.

Broad Buckler-fern (May)
Top of frond has been bitten off by deer.

TALL TUFTED FERNS: IDENTIFICATION

	First division (pinna)	Second division (pinnule)	Third division (pinnulet)	Scales on stem
Lemon-scented Fern *Oreopteris limbosperma*	Much shorter at base of frond	Parallel-sided with blunt tip; edges untoothed and slightly down-turned; lemon-like smell	Not divided to this degree	Sparse
Scaly Male-fern *Dryopteris affinis* agg.*	Shorter towards frond base; dark brown pinna base	Parallel-sided with wide, very blunt tip (with tiny teeth); sides untoothed and flat or slightly down-turned		Dense; golden to rusty brown
Male Fern *Dryopteris filix-mas*	Shorter towards base of frond	Tapers slightly to blunt, toothed tip; edges with blunt teeth, flat or slightly down-turned		Pale brown or grey brown
Mountain Male-fern *D. oreades*	Shorter towards base of frond	Tapers slightly to blunt, toothed tip; edges with blunt teeth and curled upwards		Grey-brown
Hard Shield-fern *Polystichum aculeatum*	Shorter towards base of frond; glossy; stiff	Oval; tapering from middle to a pointed tip; edges flat, with tip and teeth all ending in distinct sharp points (stiff, almost spiny points in Hard Shield-fern; softer hair-points in Soft Shield-fern)		Many; dark brown
Soft Shield-fern *P. setiferum*	Shorter towards base of frond			Many; brown or grey
Lady Fern *Athyrium filix-femina* **	Shorter towards base of frond	Most pinnules: narrow (with clear intervening gaps), tapering to rather pointed tip and with distinct teeth	Not divided to this degree	Many; brown
		Largest pinnules on some plants: deeply divided into smaller pinnulets >>>	Narrow (with clear intervening gaps); tapering to rather pointed tip; many distinct teeth	
Narrow Buckler-fern *Dryopteris carthusiana* (not v. tufted) ***	Slightly shorter towards base of frond	Larger pinnules (esp. those on larger, lower pinnae) are deeply divided into smaller pinnulets described in column to right >>> Smaller pinnules (toward tip of frond) are not subdivided further. They are blunt and parallel-sided, with pointed teeth. Their edges are flat, turned down or curled up, as described for pinnulets in column to right >>>	Blunt; parallel sides flat or slightly down-turned, with pointed teeth	Pale brown
Broad Buckler-fern *Dryopteris dilatata*	Longest at base of frond		Blunt; parallel sides down-turned, with pointed teeth	Pale brown + dark stripe down centre
Northern Buckler-fern *D. expansa*			Blunt; parallel sides flat, with pointed teeth	Pale brown
Hay-scented Buckler-fern *D. aemula*			Blunt; parallel sides curled up, with pointed teeth	Pale brown + dark base

* *Dryopteris affinis* agg. includes all three Scaly Male-ferns: *D. affinis, D. borreri* and *D. cambrensis*

** *Spore cases (on underside of frond) are of a distinct comma-like shape*

*** *Not very tufted but included here because of other similarities with these tufted species*

TALL TUFTED FERNS: HABITATS

	Habitats
Lemon-scented Fern *Oreopteris limbosperma*	On **moist acid soils** in **woods** (W11 & 17) and on **open** slopes (esp. N-facing) where it can form dense patches (U19). Of **medium palatability** to grazing animals (sheep, deer, etc).
Scaly Male-fern *Dryopteris affinis* agg.	Very common, mainly on **acid soils** in **woods** (W9-11 & 16-17), but also on **open hillsides**. Like Male Fern it is **palatable** to grazing animals such as sheep, deer and cattle.
Male Fern *Dryopteris filix-mas*	Very common, mainly on **acid soils** in **woods** (W9-11 & 16-17). **Palatable** to grazing animals such as sheep, deer and cattle.
Mountain Male-fern *Dryopteris oreades*	Among **acid to basic rocks** (ledges, screes, boulder-fields) in **uplands** from Wales northwards. **Palatable** to grazing animals.
Hard Shield-fern *Polystichum aculeatum*	On **neutral to basic soils** in woods (W8-9). Palatability unclear (mainly on steep rocky banks where access for grazing animals is very limited).
Soft Shield-fern *Polystichum setiferum*	On **neutral to basic soils** in **woods** (esp. **ancient** W8-9). Rare N of Argyll, Scottish Central Belt and the Lothians. Palatability unclear.
Lady Fern *Athyrium filix-femina*	On **acid to neutral soils**, mainly in **woodland** (W7 & W9). **Palatable** to grazing animals such as sheep, deer and cattle.
Narrow Buckler-fern *Dryopteris carthusiana* *	On **wet acid to basic soils**, in various **mires and wet woods**.
Broad Buckler-fern *Dryopteris dilatata*	Very common, mainly on **acid soils** in **woods** (W9-11 & 16-17); also in **heaths and bogs**, and on **cliffs**. **Palatable** to herbivores.
Northern Buckler-fern *Dryopteris expansa*	Among **mountain rocks** and in **wet upland woodland**; mainly in the Highlands but also S Scotland, N England and Wales.
Hay-scented Buckler-fern *Dryopteris aemula*	Mainly on **acid soils** on **steep and rocky slopes in humid woodland** (esp. **ancient** W11/17) in the west (also E to N York Moors, New Forest & The Weald)

* *Not very tufted but included here because of other similarities with these tufted species.*

Male-fern (May)

Hay-scented Buckler-fern (August)

Beech Fern and Oak Fern

Two **small** ferns with fronds on **slender stalks**, each stalk arising on its own (not grouped in tufts). Plants of sheltered and shaded habitats.

Beech Fern ►

Phegopteris connectilis has **fronds divided twice (2-pinnate)**, narrowly triangular with **pointed** tips to side pinnae and whole frond, and with the **lowest two pinnae pointing backwards**. On **moist, acid to basic soils** in **sheltered, shaded places in woods** (esp. W9/11/17) **and among rocks**. Mostly in places with **little or no grazing**. Commonest in **northern and western upland areas**.

Beech Fern in May

Oak Fern in May

◄ Oak Fern

Gymnocarpium dryopteris has **bright yellow-green fronds divided 3 times (3-pinnate)**, **broadly triangular**, with **blunt tips** to side pinnae and whole frond, and with **dark-coloured, very slender stalks**. On **moist, acid to basic soils** in **sheltered, shaded places in woods** (esp. W9/11/17) **and among rocks**; in places with **little or no grazing**. Commonest in **northern and western upland areas**.

Bracken *Pteridium aquilinum*

The **tall fronds** of Bracken reach **1–2 m in height**, have **thick, tough stems** and are **divided 3 times** (3-pinnate). They arise **singly (not in tufts)** in spring from creeping underground shoots (rhizomes). A single plant produces many fronds, forming **large, dense patches**. Fronds turn orange-brown in autumn and remain on ground through winter as mats of pale brown litter. Spore cases are on frond margins. Bracken grows on **deep, moist, acid to neutral soil** in **woods** (esp. W10–11 & 14–17) and in the **open** (U20 & W25). **Unpalatable** to grazing animals (live fronds are **toxic**). Can invade grasslands and heaths, reducing agricultural productivity and botanical diversity. Can be **controlled by cutting or spraying with chemicals**. **Trampling by cattle** can help control its spread. Cut fronds can make **compost**, or **bedding** for livestock.

Bracken in June

Bracken in October

Common Thistles - tall; prickly leaves; dense, wide, mid/dark pink flower heads

<table>
<tr><th>IDENTIFICATION</th><th>Basal rosette?</th><th>Leaf margin</th><th>Upper leaf surface</th><th>Stem</th><th>Flower diam.</th></tr>
<tr><td>Creeping Thistle
Cirsium arvense</td><td>No basal rosette</td><td>Strong spines</td><td rowspan="2">Smooth (no hairs)</td><td>No prickles</td><td>15-25 mm</td></tr>
<tr><td>Welted Thistle
Carduus crispus</td><td rowspan="4">Lower leaves form a dense basal rosette</td><td>Weak spines</td><td>Prickly</td><td>10-20 mm</td></tr>
<tr><td>Musk Thistle
Carduus nutans</td><td rowspan="2">Strong spines</td><td>Sparse non-prickly hairs</td><td>Prickly except at top</td><td>30-50 mm (drooping)</td></tr>
<tr><td>Spear Thistle
Cirsium vulgare</td><td>Prickly hairs</td><td>Prickly</td><td>30-50 mm</td></tr>
<tr><td>Marsh Thistle
Cirsium palustre</td><td>Weak spines; purplish</td><td>Non-prickly hairs</td><td>Prickly; commonly purplish</td><td>15-20 mm</td></tr>
<tr><td colspan="6">HABITATS</td></tr>
<tr><td>Creeping, Welted & Spear Thistles</td><td colspan="5">Mainly dryish neutral soil in fields, along hedges and roadsides, and on disturbed ground. (esp. MG1/6-7 & OV25). Welted T. not in N Isles or W Ireland; rare in W Scotl.</td></tr>
<tr><td>Musk Thistle</td><td colspan="5">On dry neutral to basic soils, in lowland grassland (esp. MG1 & CG2-3) and on disturbed ground. Mainly in England & Wales.</td></tr>
<tr><td>Marsh Thistle</td><td colspan="5">Mainly on moist to wet neutral soils, in woods (esp. W7) and various mires with mixtures of rushes, herbs and grasses (esp. M22-28).</td></tr>
</table>

Also: Slender Thistle *Carduus tenuiflorus*: like Welted T. but **spiny wings all the way up stem** (not quite to top in WT) + **flower heads narrow/cylindrical** (not wide/globular). **Dry soils near coasts**. V. rare Highl./W Irel.; not in N Isles.

Marsh Thistle (June)

Creeping Thistle (July): flowers paler than in the other species here

Spear Thistle: basal leaves (April) and flower (July)

Flowers develop into whitish cottony seed heads whose individual hairs are 'feathered' in *Cirsium* (L) but not in *Carduus* (R) >>>

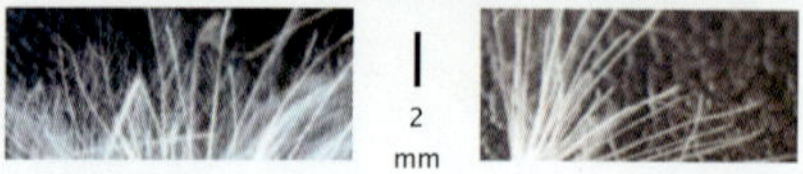

Two short Thistles of dry basic grasslands

Stemless Thistle *Cirsium acaule* has a **low basal rosette of leaves** with closely-spaced lobes and prickles. **Upperside of leaf is hairless** and differs from that of other thistles in this book in being a **brighter, shiny green. No stem**: the **pink flower head (of typical thistle-type) grows straight from the middle of the leaf rosette**. Grows in **short (grazed) unimproved dry calcareous grasslands** (esp. CG2) on **chalk and limestone in the SE half of England**; rare in SW England and Wales.

Plant in flower in July

Carline Thistle *Carlina vulgaris* is a biennial (lives for 2 years). In the 1st year there is just a **low basal rosette of leaves <10 cm long, with prickly edges and whitish cottony undersides**. In the 2nd summer these die and a stem grows up bearing similar (but shorter) leaves and topped with one or more unmistakeable **yellow-brown prickly flower heads** 2-4 cm across. Mainly in **short (grazed) or open and bare/rocky unimproved dry calcareous grasslands** (esp. CG1) at low altitudes in England, Wales and central Ireland. Scarce elsewhere; in Scotland mainly coastal.

2nd year plant in July

Flower head in autumn

Dandelion *Taraxacum officinale* agg.

An aggregate of more than 230 microspecies. **Hairless** plants with low **rosettes of unstalked oblong leaves with backward-pointing lobes**, and **large yellow flower heads** (up to 6 cm across) on **thick, hollow, leafless stems**, developing into **spherical mass of whitish fluffy seeds**. Mostly on **dry to damp neutral soils** in **gardens, roadsides, hedgerows, woodland margins** and on **disturbed (incl. trampled) ground**. Also in **flushes** (e.g. M10) and **springs** (M37).

Plant in May

Plant in April

Greater Knapweed *Centaurea scabiosa*

Branched stems up to 1.5 m tall, **sparsely hairy, deeply lobed leaves** (mostly unlobed in some plants), and **pink-purple flower heads with long, spreading outer 'rays'**. On **dry, neutral to basic soils**, mainly in **lowland unimproved grasslands** (MG1 & CG1-8) in England & Wales; scarce in Ireland; rare in Scotland. (Common Knapweed (page 276) has leaves mostly unlobed, duller green and hairier, and flowers without outer rays.)

Old flower heads in winter

Greater Knapweed

Common Knapweed

Leaves in May

Flower in June

Saw-wort *Serratula tinctoria* ▸ has **pink knapweed-like flowers** on branched stems <90 cm tall, **deeply lobed leaves** with **fine, sharp teeth** and a **large end lobe**. On **damp soil** in **unimproved grassland, heath, wetland and scrub**, in England, Wales and Galloway.

Field Scabious *Knautia arvensis*
Small Scabious *Scabiosa columbaria*

Together with the Valerians on page 189, these are the only non-woody plants in this book whose **leaves are in opposite pairs along the stems and are divided into leaflets to left and right of the leaf stalk**. In these Scabious species the stems (which can be branched) and the leaves are at least slightly hairy. The leaves at the base of the plant are **oval or lobed**, and those on the stems are **cut deeply into very narrow lobes**. The **flowers are pale mauve, very small and clustered into dense round heads**.

Field Scabious grows **up to 1 m tall**, has flower heads 3-4 cm across and has obvious **rough hairs on the stems** (hairs easy to see, and can be almost as long as the stem is wide). It grows mainly in **unimproved dry lowland calcareous grasslands** (CG3-8) and **neutral grasslands** (MG1 & 5), but also on disturbed ground. It is widespread, but is rare in N & W Scotland and W Ireland.

Small Scabious grows **up to 70 cm tall**, has flower heads 1.5-3.5 cm across and **stems hairless or with very short inconspicuous hairs**. It grows in **unimproved dry lowland calcareous grasslands** (CG1-9) in England, Wales and (rare) SE Scotland.

Cat's-ear and Hawkbits

Dandelion-like in having **yellow flowers, round fluffy white seed heads** and **low rosettes of lobed leaves**, but **leaves are hairy, lobes do not point backwards** and **flower stems solid**.

Cat's-ear *Hypochaeris radicata* has **tough-textured leaves with rough, hairy upper surfaces**, **stems branched** and **flower heads 2.5–4 cm across**. As the stems are branched there are two or more flowers per main stem (only one in Dandelion). The **leaves are held very low**, close to ground level, and their **hairs** (as seen through a hand lens) are **not branched**. Common in **short or open, heavily to lightly grazed, grassy vegetation** on **dry, neutral soils**.

Rough Hawkbit *Leontodon hispidus*: like Cat's-ear but **leaves stick up more, many of their hairs are divided part-way up into three tiny branches** (need a lens to see this) and **flower stems are unbranched** (just one flower per stem, as in Dandelion). **Flower heads 2.5–4 cm across**. Mostly in short, **grazed grasslands on dry neutral to basic soils**. Very rare N of central Scotland.

Lesser Hawkbit *Leontodon saxatilis* is similar to Cat's-ear, with leaves lying quite flat, close to the ground, but **most of their hairs are divided part-way up into two tiny branches** (need a lens to see this) and the **flower stems are not branched** (just one flower per main stem). **Flower heads small: up to 2 cm across**. Mostly in short, **grazed grasslands on dry neutral to basic soils**. Throughout England, Wales and Ireland, but scarce in Scotland.

Autumn Hawkbit *Scorzoneroides autumnalis* is delicate, with **narrow, thinly-lobed basal leaves with rather sparse hairs**. Its **stem is slender, mostly solid, can be branched and has no leaves. Flower head up to 3.5 cm across**. Common in **heavily to lightly grazed grasslands, sand dunes** & **saltmarshes**.

Smooth Hawk's-beard *Crepis capillaris* is like Autumn Hawkbit but is **hairless**, has **small, unstalked leaves on branched stems**, and has **smaller (10–15 mm) flower heads**. In **dry, open or disturbed habitats at low altitudes**.

Leaves of Cat's-ear in April

Leaves & flower of Autumn Hawkbit in August

Sow-thistles *Sonchus* species

Like a cross between a Dandelion and a Thistle, these plants have **tall, branched stems** and **oblong, lobed leaves** with **spiny teeth** along their edges. Leaves grow at the plant base and up the stems. Common in **weedy, ungrazed or little-grazed vegetation** along **hedges, roadsides, field margins, disturbed/waste (but not very trampled) ground** and (especially in the case of Perennial Sow-thistle) the **upper edges of beaches and saltmarshes**.

IDENTIFICATION OF THE THREE COMMON SPECIES OF SOW-THISTLE	**Colour of upper surfaces of leaves**	**Backward-projecting lobes (auricles) at each side of base of stem leaf**	**Hairs on stem**	**Diameter of flower head**
Smooth Sow-thistle *Sonchus oleraceus*	Dull greyish-green	Pointed	No hairs	2-3 cm
Prickly Sow-thistle *Sonchus asper*	Shiny green (not greyish)	Rounded	No hairs	2-3 cm
Perennial Sow-thistle *Sonchus arvensis*	Dull or slightly shiny green; not very greyish	Rounded	Hairy in upper part	4 cm

Perennial Sow-thistle in flower (August)

Smooth Sow-thistle stem leaves (Sept.)

Perennial Sow-thistle leaves (August)

Ragwort *Jacobaea vulgaris*

Distinctive deeply, intricately lobed leaves and stiff, ridged, branched stems to 1 m tall with many yellow flowers 15–25 mm across. Lowest leaves quite cabbage-like, with upturned edges; usually dying away before the plant flowers. Stem leaves have narrower lobes. Stems and leaves are more or less hairless. Grows mainly on dry to moist, neutral soils, in lightly to heavily grazed, short to tall grasslands (MG1, MG5, CG2–8 & U1), on waste ground, roadsides, and, more naturally, sand dunes and cliffs. A notifiable weed, poisonous to grazing animals, causing liver damage; poison can be absorbed through human skin, so it is best not to handle the plant.

Oxford Ragwort *Senecio squalidus* has leaves deeply but less intricately divided, with down-turned extreme edges. Introduced: on waste/disturbed ground and walls, especially in urban areas, and by railway lines.

Narrow-leaved Ragwort *S. inaequidens* is like Ragwort but the leaves are simple, very narrow and unlobed. Introduced. In disturbed, waste, urban and coastal habitats.

Marsh Ragwort *Jacobaea aquatica* has darker, shinier, less intricately divided leaves with large end lobes and forward-pointing side lobes, and fewer, larger flowers 25–40 mm across; on wet neutral soils, in rush/herb mires (including M23 & M28).

Groundsel *Senecio vulgaris*

Rather like Ragwort (page 168) in its **deeply lobed leaves**, but the leaves are a **greyer-green** and **can have down-turned edges**, the **stems are shorter** (to 30 cm), **weaker, and commonly reddish**, and the **flowers are narrow, cylindrical and yellow only at the tips** (flowers can also be drooping). Flowers develop into round white fluffy seed heads. Grows mainly on **dryish, neutral soils**, in **artificially disturbed (but not very grazed or trampled) habitats** such as **waste ground, roadsides, arable field margins** and **gardens**, and, more naturally, **sand dunes** and **cliffs**.

Plants in April

Heath Groundsel *S. sylvaticus* is **taller** (to 70 cm), **rather scented** (not very pleasantly so!), can have **slightly sticky hairs** on leaves, and has **flowers like Groundsel but more upright and tipped with extremely short, down-turned yellow petals**. On **dry acid to neutral soils**, mainly in **open** or **disturbed** places among **lowland heaths, grasslands and woods**.

Sticky Groundsel *S. viscosus* grows up to 60 cm tall, is **strongly smelling** (unpleasant!), has **stems and leaves grey-green with sticky-hairs**, and **flowers tipped with short down-turned yellow petals**. On **dry, acid to neutral soils**, mainly in **disturbed** habitats such as **waste ground**.

Buck's-horn Plantain *Plantago coronopus*

Low rosettes of **narrow, thick-textured leaves with small lobes or large teeth**. Variably hairy. Flowers in small brown heads. In short or sparse vegetation on **dry sandy or rocky coastal ground**. Also inland in SE Britain, especially by salt-treated roads.

Sea Plantain (page 292) differs in being hairless and lacking leaf lobes, though some of its leaves can have a very few teeth.

Sea Rocket *Cakile maritima*

A well-branched plant with **grey-green, thick-textured hairless leaves** cut into **narrow lobes**, and **small white or pale mauve 4-petalled flowers**. On **coastal sand and shingle along the High Water Mark**.

Rock Samphire *Crithmum maritimum* ► has **leaves of similar texture, divided into many long, narrow segments. Yellowish flowers in branched umbels. Coastal cliffs** (MC2) and **stony shores**, mainly from SW Scotland S & E to SE England, and around Ireland.

Poppies

These have **leaves cut deeply into rather blunt lobes**. *Papaver* species have **large red flowers** on **stiff, hairy stems**. Their **buds droop** before opening. They are **introduced** plants of **dry, neutral to basic soils**, mainly in **arable fields** and artificially disturbed (but generally **ungrazed and untrampled**) ground.

IDENTIFICATION OF COMMON RED POPPIES	Flower diameter	Petals wide & overlapping?	Capsule in middle of flower (remaining after petals fall)
Common Poppy *Papaver rhoeas*	7-10 cm	Yes	Hairless, slightly longer than wide
Long-headed Poppy *P. dubium*	3-7 cm	Yes	Hairless, at least twice as long as wide
Prickly Poppy *Roemeria argemone* *	2-6 cm	No	Hairy, about twice as long as wide
Rough Poppy *Roemeria hispida* *	2-5 cm	No	Hairy, almost spherical

* = mainly in SE England

Common Poppies in flower (June)

Common Poppy leaves (June)

Welsh Poppy (June)

Welsh Poppy *Papaver cambricum* has **yellow flowers** 5-7.5 cm across; **leaf lobes/leaflets pointed and toothed**. Mainly on **moist, neutral to basic soils in woods** (esp. W8-9) and on **cliffs**. It is native in parts of Wales, SW England and in Ireland, and **introduced** elsewhere.

Greater Celandine *Chelidonium majus* has **smaller yellow flowers** 2-2.5 cm across; **leaf lobes/leaflets wide and blunt, with very rounded teeth**. **Introduced: by roads, paths, walls, buildings etc**; widespread but rare in N/W Scotland & W Ireland. **Poisonous**.

Shepherd's-purse *Capsella bursa-pastoris*

The **low rosettes of lobed leaves** are a common and distinctive sight on bare or disturbed soil. They look a bit like dandelion leaves, but the **leaf lobes stick straight out** and do not point backwards. The **stems** grow up to about 30 cm tall and bear **smaller oblong or slightly lobed leaves** with no stalks and with **backward-pointing projections at their bases** (these projections 'clasp' the stem). The stem can have a few branches, and the upper parts of stems and branches have many very small 4-petalled **white flowers**, each one on a thin stalk and later becoming a **triangular fruit** with a tiny notch in its wide tip. **Introduced**, growing on **nutrient-rich soils** in **disturbed** places such as **arable fields, gardens and waste ground**.

Plants in May. Left = flowers and fruits. Right = Basal leaf rosette (with pale blue flowers of Common Field Speedwell in lower-left corner)

Wild Mignonette *Reseda lutea*

Up to 75 cm tall, with **leaves cut deeply into very narrow, untoothed lobes**, and stems with **spikes of small creamy flowers**. On **dry neutral to basic soils in the lowlands**, mainly in **disturbed habitats** such as roadsides and waste ground, and in **unimproved dry grassland** (e.g. CG2-3 & U1). Most common in England; very rare in N Scotland and (introduced) Ireland.

Plants in June. Left = flowers. Right = looking down into foliage.

Louseworts *Pedicularis* species

Hairless or almost hairless herbs whose **stems are commonly purplish** and whose **leaves are 1-4 cm long, narrow oblong-shaped and deeply divided into many closely-spaced small, toothed lobes**. The leaves are commonly tinged purplish. **Purplish-pink flowers** are rather clustered toward shoot tips. Both species are **commonest in upland areas**.

Lousewort *P. sylvatica* ▶ is **short** (mostly 5-20 cm tall) and branched, with **all branches arising from the base of the plant**. The leaves are <3 cm long. On **damp, acid soils**; mainly in **wet heaths** (M14-16, H4-8) and **grasslands** (U4-6). **Not very palatable** to animals such as sheep and deer. **Commonest in grazed places**; grazing keeps vegetation short, preventing other species from growing tall and smothering small plants such as Lousewort.

1 cm

1 cm

◀ **Marsh Lousewort** *P. palustris* is **taller** (up to 60 cm) with leaves up to 4 cm long and **branches arising from a single main upright stem**. On **damp or wet, neutral to basic soils**, in **sedge-rush-herb mires** (mainly M9, M13, M15, M22-23 & M27).

Yarrow *Achillea millefolium*

A herb <40 cm tall with **narrow, darkish-green leaves deeply and intricately cut into short, very thin lobes**, the main lobes divided further into smaller ones (2-pinnate), which themselves can be divided (3-pinnate) or toothed. The whole leaf is narrower and with a clearer outline than that of Mayweeds and Pineappleweed (page 175). Stems quite stiff, with **dense, branched heads of small white or pale pink flowers** in summer or autumn. On **dry to moist, basic to mildly acidic soils, in unimproved or semi-improved grasslands** in lowlands and uplands, and on **sand dunes** and **disturbed ground**.

Sneezewort *A. ptarmica* has similar branched flower heads on stems up to 60 cm tall, but the individual **white flowers** are **larger** and **fewer**, and the **dark green leaves** are **simple, narrow** and pointed, with **many small sharp marginal teeth**. The leaf shape/arrangement is that of Section 3e (page 253). On **damp to wet neutral soils**, in **sedge/rush/herb mires** (e.g. M22/23/27/28).

Sneezewort leaf ▸

Mayweeds and Pineappleweed

Low-grown **hairless** herbs with **leaves deeply and intricately cut into very thin lobes**, the main lobes divided further into smaller ones (2-pinnate) that can themselves be divided (3-pinnate). Mainly on **dryish fertile soils in open, disturbed habitats** in the lowlands.

Scentless Mayweed *Tripleurospermum inodorum* has **leaf lobes pointed, and flower heads 30–45 mm across with yellow centres surrounded by long white petals. No smell.**

Sea Mayweed *T. maritimum* is **similar but slightly fleshy-textured and with leaf lobes blunt-ended.** Along **seashores and coastal cliffs.** Again, **no smell.**

Scented Mayweed *Matricaria recutita* is like Scentless Mayweed but with **a smell (not pineapple-ish)** and with **smaller flowers 10–25 mm across.**

Pineappleweed *M. discoidea* is like Scentless Mayweed but has a **strong pineapple smell and its flowers have no white petals. Tolerant of trampling**.

Scentless Mayweed in April

Pineappleweed in April

Pignut *Conopodium majus*

A delicate, hairless plant up to 40 cm tall, with **leaves deeply cut into narrow lobes/leaflets**. Lowest leaves are divided, then sub-divided and then sub-divided again to give a detailed, lace-like pattern. Stem leaves are less intricately divided. **Umbrella-shaped 'umbels' of small white flowers** in May-June. On **dry to moist, acidic to slightly basic soils**, **in woods** (esp. W11), **hedgerows** and **unimproved, grazed or mown grasslands** (esp. U4 & MG5). The final leaf divisions are not quite as thin as in Yarrow, Mayweeds and Pineappleweeds (pages 174-175). Burnet-saxifrage (page 187) looks similar when in flower, but its basal leaves are not double- or treble-divided (they have fewer, broader leaflets) and it grows on more basic soils.

Above: flowers and stem leaves in May. Below: basal leaves in April

Hedgerow umbellifers with finely-divided leaves

These four species are **tall**, at least slightly **hairy**, with **triangular leaves with many leaflets**, and **umbrella-shaped** clusters of small **white flowers**. They grow mainly on **dryish, neutral soils**, along lush, **little-grazed hedgerows** (MG1), **roadsides, wood margins** and **riverbanks**. Fern fronds look a bit like umbellifer leaves but grow only from the base of the plant (each one on a stiff scaly stalk). Ferns also lack flowers.

Upper leaf surface hairless or with short hairs sticking up. Leaflets pointed. Flowers mainly **April–June** (the earliest flowerer).

Like Cow Parsley with **pointed leaflets**, but **dense & bushy, aniseed-scented**, with **pale blotches common on upperside of leaf**. Flowers **May–June**. **Not native**.

Upper leaf surface with short, forward-pointing hairs. Leaflets rather blunt. **Stem purplish**. Flowers **June–July**.

Upper leaf surface with short, forward-pointing hairs. Leaflets pointed. Flowers **July–August** (the latest flowerer).

Three more umbellifers with finely-cut leaves

Hemlock *Conium maculatum* is **tall** (to 2 m or more) and **white-flowered** as in the species on page 177, but is **hairless** and its **stems (including the lower parts of the leaf stalks) are green with purple spots**. **Introduced**, on **dry or damp, neutral soils** in **hedgerows**, along **rivers and ditches**, and on **disturbed or waste ground** in the lowlands. **Poisonous**.

Hemlock Water-dropwort *Oenanthe crocata* is similar and hairless but its **stems lack purple spots**, the **smallest leaf segments are not quite as narrow** and the plant can be denser and rather bushy. It grows up to 1.5 m tall. On **wet, neutral soils** at low altitudes, **by water**, in **herb-rich mires/fens** (esp. M27-28, S26 & S28), **swamps** (S3-4, S6 & S14) and in **wet woods** (esp. W7). Commonest in the **west**. **Poisonous**.

Wild Carrot *Daucus carota* is **hairy**, with **stiff, rough-textured stems** as in Rough Chervil and Hedge Parsley (page 177). Stems **lack the purple tinge** of Rough Chervil. **Base of leaf stalk has long hairs** (also present in Rough Chervil but not in Hedge Parsley). The **underside of the flower head has long, projecting, branched, leaf-like bracts** sticking out all around. On **dry, basic soils in lowlands**, esp. in **grasslands** (CG1-8), on **sea cliffs** (MC4-5, MC8-9 & MC11), in **sand dunes** (SD8) and **hedgerows**, and on **disturbed/waste ground**.

Hemlock leaf (April)

Hemlock Water-dropwort leaf (January)

Hemlock stem

Wild Carrot (June)

Hogweed *Heracleum sphondylium*

Tall (to 2 m). **Whole plant hairy and rough-textured**. Stem **thick, stiff and ridged**. Leaf **large, hairy, dull green, to 60 cm long.** Leaf stalk **thick and green or purplish** with a swollen sheathing base. Leaflets **with wavy or lobed (+toothed) edges;** lowest pair stalked. **Large flower heads** (to 15 cm across) with small **white or cream flowers** (Jun–Sep). In lush **ungrazed/lightly grazed** vegetation on **neutral soils**, e.g. **grassland** (MG1), **bramble** (W24), **hedgerows, roadsides, wood edges** & **riverbanks**.

Plants in June

Giant Hogweed *H. mantegazzianum* is similar but bigger: stems to **3.5 m tall** and 10 cm wide with **red spots**; **leaves to 1 m long**; **flower heads to 50 cm across**. **Introduced**. On **disturbed ground** & by **rivers**, in **lowlands**. **Touching the stems can cause skin problems**.

Parsnip *Pastinaca sativa*. Like Hogweed but more branched, **leaves yellower green**, leaflets more numerous and **less divided** (+ **lowest unstalked**); umbels smaller, more numerous, with **yellow flowers**. **Dry calcareous–neutral soils in lowlands**; native in England N to Newcastle and in S Wales; non-native elsewhere in Britain/Ireland). **Touching stems can cause skin problems**.

Angelica *Angelica sylvestris*

Another large umbellifer. **Hairless,** with **thick, purple-tinged stems** up to 2 m tall.The **large, triangular leaves** are divided into leaflets with finely toothed edges; they are less intricately divided than those of Cow Parsley, Hedge-parsley, Rough Chervil and Sweet Cicely, and lack the thick, stiff, scaly central stem of fern fronds. The **leaf stalks are least partly tinged purple**; those growing up the stems have swollen bases. The **flower heads are large (up to 15 cm across) and domed**. Grows mainly on **moist to wet, neutral soils**, in lush, **moderately to lightly grazed herb-rich mires and fens** (M22-28) and on damp **cliff ledges** (U17).

Flowering stem in July

Lower leaf in June

Ground Elder *Aegopodium podagraria*

Like Angelica (page 180) this is a **hairless umbellifer** with **triangular leaves divided into oval leaflets** with pointed tips and finely-toothed edges. Smaller than Angelica, with green (not purple-tinged) stems and leaf stalks, and fewer leaflets (9 on most leaves). Leaves can form low, dense, extensive carpets. Flowering stems are up to about 1 m tall, with domed clusters of small white flowers. **Introduced**. Mainly on **moist, nutrient-rich soils** where there is some **shade** and **disturbance**: **hedgerows, woodland margins, roadsides, waste ground, gardens and near buildings**.

Leaves in May

Flowering plant in June

Alexanders *Smyrnium olusatrum*

Rather like Ground Elder (see above) and Angelica (page 180) but has **glossy leaves** and **pale yellowish-green flowers**. Up to 1.5 m tall. Mainly on more or less neutral soils along **hedgerows, roadsides, paths** and **walls**, and on **cliffs**; commonest near **coasts**. Scarce in Scotland. **Introduced**.

Silverweed *Potentilla anserina*

A **creeping** plant with **silvery-green leaves with silvery-white undersides**, and 5-petalled **yellow flowers**. On **dry to damp, neutral soils**, in **open short grassland** (MG11), **dune slacks** (esp. SD17), upper edges of **saltmarshes and seashores**, by **roadsides** and on **waste ground**. **Unpalatable** to grazing animals such as cattle.

Flowering plant in May

Common Stork's-bill *Erodium cicutarium*

Branched, hairy stems commonly <20 cm tall but can reach 60 cm. Stems can be reddish. **Leaves with 4-7 pairs of leaflets with edges deeply toothed** (cut more than halfway to central vein of leaflet). **Pale pink flowers 12-14 mm across, with 5 petals. Fruit with long 'stork's bill'.**

Flowering plant in October

Rather like Herb-Robert (page 206) but with leaflets in left & right rows instead of radiating from central point, and petal tips not notched. On **dry neutral to basic (including sandy) soils** at low altitudes, in **unimproved short or open grasslands** and **dunes**.

Musk Stork's-bill *E. moschatum* is similar but more hairy and with **edges of leaflets cut less than halfway to central vein of leaflet**. In similar habitats but **introduced** and less common (and very rare N of Cumbria & Yorkshire and in NW half of Ireland).

Sea Stork's-bill *E. maritimum* is smaller, with **basal part of leaf (near leaf stalk) cut into 1-2 pairs of wide, overlapping lobes, and rest of leaf with broad, blunt teeth and blunt tip**. In **short/sparse coastal dune & grassland**: Galloway S & E to Kent, & in S & E Ireland.

Vetches

Vetches have leaves divided into **small, narrow leaflets** and in most species ending in **tendrils** (curly, wiry extensions of leaf stalks). **Two stipules** (like miniature leaves) at base of each leaf stalk. Most species are **scramblers**.

Bush Vetch *Vicia sepium*: **5–9** pairs of leaflets, each of which is widest below the middle (i.e. about 1/3 of the way up from the leaf base; the edges then taper in gradually from here to the tip); **untoothed stipules**; short spikes of **2–6 dull purple flowers**. Mainly on **dry to moist, neutral soils** in **woods** (esp. W8–11) and **hedges**.

Common Vetch *Vicia sativa*: **4–8** pairs of leaflets, each of which is widest around or beyond the middle (unlike the situation in Bush Vetch); **stipules commonly toothed** and with a dark blotch; **pink-purple flowers singly or in pairs**. On **dry, neutral soils** in **unimproved grasslands** (MG1, MG3 & MG5).

Tufted Vetch *Vicia cracca*: **6–12** pairs of leaflets, each of which is widest about halfway along (edges are parallel for much of the leaflet's length); **untoothed stipules**; **long spikes of 10–40 purple flowers**. On **dry to damp, neutral soils** in **hedges, tall grassland** (MG1/5 & CG3-6) & **fens/mires** (M22–23, 27–28, & S24–26).

Bitter-vetch *Lathyrus linifolius*: **non-scrambling**; **2–4** pairs of leaflets, each one widest around middle; **no tendrils**; **stipules toothed**; short spikes of **2–6 pinkish flowers** commonly turning bluish with age. On **dry, acid to neutral soils** in **woods** (esp. W11), **heaths** (H12 & H16) and **unimproved grasslands** (U4).

Meadow Vetchling *Lathyrus pratensis*: **greyer**-green; **one** pair of leaflets, each one widest around middle; **stipules as big as leaflets**; **long-stalked** spikes of **5–12 yellow flowers**. Mainly on **dry, neutral soils** in **unimproved grasslands**.

Bush Vetch (June)

Meadow Vetchling (June)

Tufted Vetch (June)

Kidney Vetch *Anthyllis vulneraria*: **round yellow flower heads** to 1.5 cm across; leaflets with **whitish edges & undersides**; end one large. **No tendrils**. Low-grown. **Dry neutral/basic grassl. & coastal habitats**.

Fumitories and Climbing Corydalis

Scrambling or climbing plants with **finely-divided leaves**. Smallest leaf segments typically joined together as **lobes (mostly in threes) in Fumitories** but **separated into single oval leaflets in Climbing Corydalis, whose leaves commonly end in branched tendrils** (like curly extensions of leaf stalks). Leaves greyish-green (yellower-green in White Ramping-fumitory). **Spikes of small flowers: pink-purple in Fumitories; cream in Climbing Corydalis.** Fumitories differ from each other in small flower details, but other ('easier') features of the commoner species are given below.

IDENTIFICATION and HABITATS	Leaflet at end of leaf		Max. height of plant	Main habitat (note - Fumitories grow mainly in lowland areas)
	Width	Length		
White Ramping-fumitory *Fumaria capreolata*	>2.5 mm	8 mm	100 cm	Hedgebanks/scrub
Common Ramping-fumitory *Fumaria muralis*	>2.5 mm	2-6 mm	50 cm	Arable farmland + Hedgebanks/scrub (mainly acid soils)
Common Fumitory *Fumaria officinalis*	1.2-2.2mm	3-3.5 mm	20 cm	Arable farmland (mainly basic soils). Introduced.
Tall Ramping-fumitory *Fumaria bastardii*	1.5 mm	6 mm	30 cm	Arable/horticultural land (mainly in W, on acid soils)
Dense-flowered Fumitory *Fumaria densiflora*	<1 mm	2-4 mm	20 cm	Arable land (mainly in E, on basic soils). Introduced.
Climbing Corydalis *Ceratocapnos claviculata*	<2.5 mm	5-12 mm	80 cm	Woods, scrub & bracken; acid soils; rare in Ireland

1 cm

Common Fumitory

Climbing Corydalis

Bitter-cresses *Cardamine* species

Short plants; leaves with **many small oval or roundish leaflets** with no teeth or a few blunt teeth. **Small 4-petalled white or pale pink flowers at stem & branch tips**: long thin green pods later. **Leaves edible** but those of Large Bitter-cress are unpleasant.

IDENTIFICATION	Lower leaves	Stems	Flowers
Cuckooflower / Lady's Smock *C. pratensis*	forming a rosette; with 1-7 pairs of dull green or purplish-green leaflets (each 1-1.5 cm long x <1 cm wide)	to 60 cm tall; 1-2 mm thick; smooth; with 2-4 leaves	pale pink; 10-15 mm across; 4 stamens
Hairy Bitter-cress *C. hirsuta*	in a rosette; with 1-3 pairs of dull green leaflets (each <1 cm long x <1 cm wide)	to 20 cm tall; 1 mm thick; smooth; with <4 leaves	white; <5 mm across; 4 stamens
Wavy Bitter-cress *C. flexuosa*	in a rosette; with 3-6 pairs of dull green leaflets (each <1 cm long x <1 cm wide)	to 40 cm tall; 1-2 mm thick; grooved; with >4 leaves	white; <5 mm across; 6 stamens
Large Bitter-cress *C. amara*	not in a rosette; with 2-4 pairs of bright green leaflets (each 1-2 cm long x up to 1.5 cm wide)	to 60 cm tall; 1-2 mm thick; smooth; with >4 leaves	white; 12-15 mm across; 6 stamens

HABITATS

Cuckooflower	**Wet neutral soils**, in **grasslands and mires** (MG9-10, M22-28/37-38).
Hairy B-c.	**Dry neutral to basic soils** in **open, rocky or disturbed places** including **gardens**.
Wavy B-c.	**Moist neutral soil**: **mires** (MG10, M22-28); **woods** (W7-11); **by streams**; **disturbed ground**.
Large B-c.	**Moist neutral soils**, in **mires** (MG10, M22-28), **woods** (W1-7) and **streamsides**. Mainly in lowlands of England (except SW), Scotland & N Ireland.

Cuckooflower leaves & flowers (May)

Large Bitter-cress (May)

Hairy Bitter-cress in June

Wavy Bitter-cress in May

Meadowsweet *Filipendula ulmaria*

A tall, hairless plant whose leaves have 2-5 pairs of **toothed leaflets interspersed with much smaller leaflets along the dark reddish leaf stalks.** The end leaflet is 3-lobed. A pair of toothed stipules (like small leaflets) grow at the base of the leaf stalk. The tall stems (to 1.2 m) are commonly reddish and in summer end in branched **clusters** of small, scented, **creamy-white flowers 2-5 mm across**. On **moist to wet, neutral soils** in **woods** (esp. W3-7), **mires and fens** (M22-28), **damp neutral grasslands** (MG1-3 & 8-10), among **bracken** (W25) and on **cliffs** (U17). **Palatable** to grazing animals such as sheep and cattle, but low, non-flowering plants can survive in short, grazed vegetation.

Meadowsweet plants in July

Meadowsweet leaf in May

Dropwort *F. vulgaris*
is similar but the **flowers are larger (5-9 mm across)** and the **leaflets are much narrower and oblong with deeply/sharply toothed edges.** In **unimproved dry lowland calcareous grassland** (CG1-6) in England; rare in Wales, S. Scotland and W. Ireland.

Burnets and Burnet-saxifrage

Salad Burnet *Poterium sanguisorba* is hairless or sparsely hairy, up to 50 cm tall (shorter in heavily grazed grassland), has leaves with 4-12 pairs of **sharply-toothed, short-stalked, round to oval leaflets** largest toward the leaf tips, and **small green to pinkish-green flowers** in dense, roundish clusters at the shoot tips. Grows on **dry, basic soils** in **unimproved calcareous grassland** (CG1-8) in **English and Welsh lowlands**. Rare in southern Scotland and in Ireland.

Great Burnet *S. officinalis* is **taller** (up to 1 m) and hairless with **leaflets heart-shaped at their bases** and **flower heads dark red**. In **damp, neutral, unimproved grasslands** (esp. MG3-5 & M26) in **England & Wales** (rare in S Scotland & N Ireland).

Salad Burnet leaves (June)

Salad Burnet flowers (June)

Great Burnet leaf (July)

Great Burnet flowers (July)

Burnet-saxifrage *Pimpinella saxifraga*: **leaves like Salad Burnet but leaflets unstalked**. Leaves on stems cut into very thin segments. Stems 30-60 cm tall. **Small white flowers** in **clusters 2-5 cm across** as in Pignut (page 176) whose lower leaves are much more divided. On **dry, basic soils**, in **unimproved calcareous grassland** (CG1-10). Widespread. Scarce in NW Scotl. and W Irel..

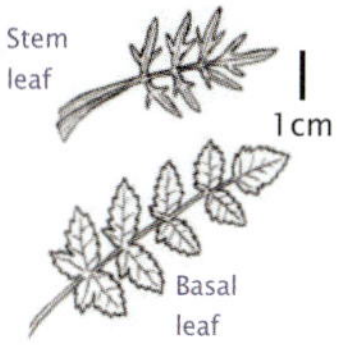

Marsh Cinquefoil *Comarum palustre*

Easily noticed by the **leaves, most of which have 5 leaflets with toothed edges and blunt tips**. Leaves hairless or slightly hairy; **bluish-green in summer**; yellower in autumn; can form low carpets; some have one or more additional smaller leaflets lower down the leaf stalk. **Dark purple flowers** on branched stems (to 40 cm tall) in summer. On **wet, mainly neutral to slightly basic soils** in **mires/swamps** (esp. M4-5, M9, S9-11 & S27). Commonly with Bogbean (page 197), Bottle Sedge (page 124) and Water Horsetail (page 220).

Agrimony *Agrimonia eupatoria*

Up to 1 m tall with **spikes of small yellow flowers**. Stems hairy.Leaves **hairy**, with **large + small leaflets** and **green stalks**. On **dry, neutral to basic soils**; in **unimproved** lowland **grassland** (MG1, MG5 and CG2-7) and in **hedgerows**.

Agrimony fruit: lower part grooved to its base; all bristles point forwards

Fragrant Agrimony fruit: grooves not extending to base; lower bristles point out and slightly backwards

Fragrant Agrimony *A. procera* is similar but has **scented leaves, notched petal tips** and **different fruits**. Grows in similar habitats to Agrimony but is less common.

Valerians *Valeriana* species

Together with the Scabious species on page 165, these are the only non-woody plants in this book whose **leaves are in opposite pairs along the stems and are divided into leaflets or lobes on their left and right sides**. Valerians are mostly hairless. Their stems, which can be red-tinged, have branched **clusters** of small **pale pink flowers**. **Palatable** to grazing animals. On **moist to wet, neutral soils**: **Common Valerian** *V. officinalis* in **woods** (esp. W3-7), **mires and fens** (M22-28), **damp neutral grasslands** (MG1-3 & 8-10) and on **cliff ledges** (U17); **Marsh Valerian** *V. dioica* in **mires** (M22-27) in England, Wales and south-easternmost part of Scotland.

Common Valerian (July)

Common Valerian: **up to 1.5 m tall**; each leaf has 3-6 pairs of **bluntly toothed leaflets** and another leaflet at the end.

Marsh Valerian ▸ **up to 30 cm tall**; less branched; **flowers in spring**.

Wood Avens & Water Avens *Geum* species

Medium-height plants to about 50 cm tall, with hairy, branched stems and hairy leaves. **Lowest leaves have 3 large leaflets and 1–5 pairs of much smaller leaflets** lower down the leaf stalk. All leaflets have toothed edges. Leaves up stems are smaller, **undivided or with 3 leaflets**, with a pair of **stipules** (like small leaves) where the leaf stalk joins the stem. Both species grow mainly on **neutral to basic soils**.

Wood Avens *Geum urbanum* has **leaves present all year round**; up to 4 pairs of leaflets on lower leaves; **end leaflet** 1- or 3-lobed, **at least as long as wide**, and commonly tapering gradually to its base. **Small yellow flowers** with 5 petals; rather like small buttercup flowers with narrower petals. On **dry to moist, neutral to basic soils** in **woods** (W8–9), especially where there is **little or no grazing**.

1 cm

Flower

1 cm

Upper leaf & stipules

1 cm

1 cm

1 cm

Hybrids are locally common; their flowers are intermediate between those of the two species.

Water Avens *Geum rivale* has slightly darker, more toothed leaves whose leaflets are more folded along their central and side veins. Leaves scarce in winter. Lowest leaves have **3–6 pairs** of leaflets; **end leaflet** 1- or 3-lobed, **at least as wide as long**, and typically tapering abruptly to base. **Drooping pinkish flowers**. On **damp, neutral soils** in **woods** (esp. W7), **mires** (esp. M10–11, 23 & 27), **grasslands** (esp. CG10, MG3 & herb-rich forms of U4–5) and **cliff ledges** (U17). Quite tall where ungrazed or lightly grazed but can survive as low rosettes of leaves where more heavily grazed.

1 cm

1 cm

◄ **Some winter basal leaves of Wood Avens are simpler**: roundish or heart-shaped; edges with teeth or 3 shallow rounded, toothed lobes; like leaves of Ground Ivy (page 222, which have fewer, larger, more rounded teeth) and Garlic Mustard (page 256, which are hairless).

Section 3b

LEAVES WITH LOBES OR LEAFLETS RADIATING OUT FROM CENTRE (PALMATELY DIVIDED)

This section begins with species whose leaves have just:

THREE LEAFLETS as in upper leaf in drawing on right: see pages 192-198...

...before moving on to those whose leaves are more complex, with:

THREE OR MORE LEAFLETS OR LOBES (e.g. lower leaf in drawing on right): pages 198-211.

Within each of these two groups, species can be identified by such things as:

- presence or absence of teeth along the edges of lobes/leaflets
- hairiness
- leaf colour
- details of leaf stalks
- and, of course, flowers

Note: some leaves of Creeping Buttercup (page 200) can look more fitting for section 3a (leaves with lobes or leaflets in two rows either side of a central leaf stalk) because they have 5 lobes: two opposite pairs + one end lobe (see photo ►).

Red and White Clovers

White Clover *Trifolium repens.* Mostly < 20 cm tall. Leaves divided into **three very blunt, broad oval or round leaflets**. Each leaflet has **many straight veins** on each side of its central vein (unlike Wood Sorrel - see page 194 - which has a superficially similar-shaped leaf), and a **pale blotch or V-shaped or crescent-shaped line** across its middle (absent in Wood Sorrel). **White flowers in dense round or oval clusters** at tips of leafless flower stems. On **dry to moist, neutral soils** in **grasslands**, especially where **improved or semi-improved and nitrogen-rich** (MG6-7, U4b). Also in some unimproved grasslands (MG5, CG10).

Red Clover *T. pratense* has **relatively longer, narrower and more pointed leaflets** of a greyer-green colour, and **larger heads of pink flowers** on **stiffer stems** that are straight and have leaves on them. Can grow to 50-60 cm tall. Mostly on **dry to moist, neutral soils**, in **unimproved grasslands** (mainly MG5) and in **hedgerows**.

White Clover

Red Clover

Zigzag Clover *T. medium*: like Red Clover but has **narrower leaflets** with **pale markings faint or absent**. Stipules (small leaf-like bracts at base of leaf stalk) **taper gradually to their tips** (in Red Clover tapering less gradually to hair-like bristle points). Can form **dense patches**. **Neutral to acid grassland, heath** and **woodland edges**. Less common than Red Clover.

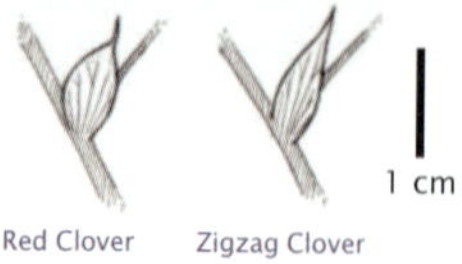

Stipules

Alsike Clover *T. hybridum*: growth form like Red Clover; leaves like Red or White Clovers but no pale markings; **flower head a mix of white (esp. upper part) and pink (esp. lower part)**. **Introduced**. Mainly in lowlands, in **grasslands** and on **disturbed/waste ground**.

Yellow-flowered clover-like species

These are short plants with **leaves like those of clovers** (page 192) but with **smaller, narrower leaflets** whose tips are blunt or notched (some with a tiny projection at the tip). They have **small roundish clusters of yellow flowers**.

Lesser Trefoil *Trifolium dubium* (lower-left photo) has **leaflets <5 mm wide and up to 8 mm long**, with blunt or notched tips with **no projection or just a very narrow indistinct hair-like projection** at the extreme tip. The **stems are hairless or sparsely hairy**. **Flower heads small (up to 6 mm long) and not very dense.** In a wide range of **short grasslands** (**meadows or pastures**) on **dry to damp, acid, neutral or basic soils** at low altitudes; also on barer soily or rocky ground.

Hop Trefoil *Trifolium campestre* (photo at right) is similar but slightly larger, with **leaflets >5 mm wide and up to 10 mm long, and larger, denser flower heads 10–15 mm long** (these can have a 'puffy' or 'inflated' look); **stems hairy**. Mainly in **short grasslands** (**meadows/pastures**) on **dry, neutral to basic soils** in the **lowlands**.

Black Medick *Medicago lupulina* (lower-right photo) is also larger than Lesser Trefoil, with **leaflets 6–15 mm wide and up to 20 mm long** with a **tiny triangular point projecting at the tip**, and **flower heads up to 8 mm long** (denser than in Lesser T. but not inflated-looking as in Hop Trefoil). **Stems hairy**. Mainly in **short grasslands** (**meadows or pastures**) on **dry, neutral to basic soils** in the **lowlands**.

Wood Sorrel *Oxalis acetosella*

A short plant whose leaves are divided into **three heart-shaped, untoothed leaflets** as in Clovers but each leaflet **folded** upwards to either side of its central vein, and with **about three indistinct curved veins** to left and right of the central vein. (Clover leaflets are pointed or slightly heart-shaped, flatter and with many straight side-veins; some also have teeth and/or a pale central blotch.) Mainly on **dry to moist, acid soils** in **shaded places**, especially **woodland** (W9–W11). Also under **bracken** (U20), in **hedgerows** and **heaths**, and among **rocks**. In southern lowlands mostly in **ancient** woods. **Leaves are edible** (acid tasting) but **not very palatable** to grazing animals such as sheep.

Plants in flower in May

Wood Sorrel leaf

White Clover leaf (for comparison)

Bird's-foot-trefoils *Lotus* species

Common Bird's-foot-trefoil *Lotus corniculatus*. Short, with leaves divided into 3 small (<1 cm long) oval, greyish-green, untoothed and more or less hairless leaflets and two similar-looking 'leaves' (called stipules) at the base of the leaf stalk (this stalk being up to 6 mm long). In summer the yellow or yellow-orange pea-shaped flowers are conspicuous; later they become thin, dark-coloured seed pods up to 3 cm long. Grows on dry, neutral to basic soils, mainly in short, grazed, unimproved grasslands (CG1-10 & 13-14, MG1 & 3, and U1 & 4) and also sand dunes, and some dwarf shrub heaths (H6-8 & 10). Where acid grasslands predominate this species can, when in flower, be one of the more conspicuous indicators of localised slight base-enrichment.

Common Bird's-foot Trefoil in July

Greater Bird's-foot-trefoil *L. pedunculatus* is taller (to 50 cm or more - can scramble up among other plants in tall vegetation), with hairy, larger leaflets (>15 mm long; leaf stalk up to 10 mm long) and hollow stems (solid in Common Bird's-foot-trefoil). Side veins of leaves clearly visible as paler translucent lines when held up to the light; Common BFT side veins are barely apparent even when held up to light. Common on damp or wet neutral soils, in rush and herb mires/fens (M23-28).

Narrow-leaved Bird's-foot-trefoil *L. tenuis* is like Common BFT but leaflets narrow (end leaflet <5 mm wide and with width less than 1/3 of its length). In similar habitats to Common BFT but scarcer: mainly in lowlands of SE Britain; rare elsewhere.

Slender BFT *L. angustissimus* is like Narrow-leaved BFT but leaves <1 cm long; flowers in groups of only 1-2. On dry soils in open habitats in S England; mainly coastal.

Hairy BFT *L. subbiflorus* is like a small Common BFT but its stems and leaves are densely hairy. On dry/thin soils in open coastal habitats in S-SW England, SW Wales & S Ireland.

Strawberries and Stone Bramble

Plant in May

Wild Strawberry *Fragaria vesca.* Leaves long-stalked with three rather sharply-toothed leaflets (tooth at tip at least as large as neighbouring teeth), their uppersides bright green with prominent veins and undersides pale green with hairs closely pressed (can look quite smooth). Small white 5-petalled flowers in spring and summer, later becoming small edible red strawberries. Spreads by above-ground shoots (runners). Common on dry to moist, neutral to basic soils in woods (esp. W8-10), scrub (W21), hedgerows & rocky habitats.

Plant in April (leaves still small)

Barren Strawberry *Potentilla sterilis.* Similar but leaves mostly smaller and duller green, with fewer, blunter teeth (tooth at tip smaller than neighbouring teeth) and more conspicuous hairs sticking out on upper and lower leaf surfaces (not closely pressed as in Wild Strawberry). No fruit. Common in similar habitats but less confined to shady places.

Stone Bramble *Rubus saxatilis* has leaves like those of Wild Strawberry but the end leaflet is stalked. Stems creep above ground. White 5-petalled flowers like those of Bramble (page 73), but smaller. Fruits like Bramble but red, with fewer individual berries. On neutral to basic soil in rocky woods (W9) & open habitats in upland areas, mainly in N Britain.

Flowering in June

Fruiting in August

Bogbean *Menyanthes trifoliata*

Unmistakeable **thick-textured, hairless, slightly greyish-green, oval, untoothed leaflets on thick upright leaf stalks**. Leaves can form low carpets. Blunt **spikes of white flowers on upright stems** (up to 30 cm tall) in summer. On **very wet, acid to neutral soils** (especially where peaty) in **sedge mires** (esp. M4-5), **blanket bogs** (M17-18), **valley bogs** (M21), **vegetated bog pools with shallow water** (M1-2), **larger, deeper, less vegetated bog pools**, and in **swamps/fens** (esp. S4, S9-11 & S27). Commonly with Marsh Cinquefoil (page 188), Bottle Sedge (page 124) and Water Horsetail (page 220).

Leaves

Flowers (July)

Growing among *Sphagnum* mosses (view looking down from above)

Tormentil *Potentilla erecta*

A short plant whose small, mid to darkish green leaves are divided into 3 leaflets. Basal leaves are stalked. Those on stems are **unstalked and look 5-lobed** because there are **two leaflet-like stipules** at each leaf base. Leaflets and stipules are up to 2 cm long and **narrow oval with a few relatively large teeth**. **Small yellow 4-petalled flowers** in summer. Common on **dry to damp acid soils** in **unimproved grasslands** (U2-7), **heaths** (H7-8/10/12-22 and M15), **woodland & scrub** (W4/11/17-19) and in **bracken** (U20). Also on neutral to basic upland grasslands (MG5 & CG10-12) and *Dryas* heaths (CG13-14). Commonly with Heath Bedstraw in **short**, heavily **grazed** upland acid grasslands and heaths: both species are **not very palatable** to grazing animals.

Trailing Tormentil *P. anglica* is similar but **all leaves stalked**, with **3-5 leaflets** (only 3 in Tormentil) and flowers have **4-5** petals (4 in Tormentil). Mainly on **dry acid soil** in low altitude **woodland edges/clearings** and **hedgebanks**. Scarce N of Galloway.

Creeping Cinquefoil *P. reptans*: **larger**, with **long creeping stems** (stolons), **stalked leaves with 5-7 leaflets** up to 3 cm long (5+ teeth on each side); **flowers (summer) have 5 petals**. On **dry, neutral to basic soils** in **hedgerow, roadside, woodland edge** and **disturbed/waste ground** habitats. Mainly in lowlands. Scarce N of Central Scotl.

Spring Cinquefoil *P. verna*: **leaves with 5-7 leaflets** (<5 teeth per side; end tooth small); **5-petalled flowers in spring**; in **unimproved dry calcareous grassland** (e.g. CG7) and **rocky habitats**; mainly lowland; widespread but very patchy distribution in Britain; unrecorded in Ireland.

Alpine Cinquefoil *P. crantzii*: like Spring C. but leaflet has **'normal-size' end tooth**; **larger flowers >15mm across** (<15mm in Spring C.), in **summer**. **Montane calcareous grassland/cliffs** (CG10-12, U17) in Scottish Highlands, N. Pennines and (rare) Southern Uplands, Lake District, Peak District and N Wales.

Hoary Cinquefoil *P. argentea* has 5 **narrow, deeply-toothed leaflets with silvery hairs (esp. below)**; 5-petalled flowers in summer; **dry grassland** (e.g. U1) & **waste ground** in British lowlands, esp. SE England.

Wood Anemone *Anemone nemorosa*

Each leaf is stalked and has 3 leaflets, each one deeply divided or toothed; the side leaflets can be divided almost to their bases, so the whole leaf then appears to have 4–5 leaflets. Each flowering stem has three leaves all arising from the same place, spreading in different directions. After flowering, more leaves grow singly (not in threes) from a creeping underground stem (rhizome). The white, 6-petalled flowers are conspicuous in spring. The plant is hairless or with sparse hairs. It grows on dry to moist, acid to neutral soils in woods (esp. W7–11). Can form arpets. Also in heaths (esp. H12/21) and unimproved grasslands (esp. U4), mainly on N-facing slopes. In southern lowlands mostly in ancient woods.

Plants in May

Rue-leaved Saxifrage *Saxifraga tridactylites*

A small, hairy annual up to 10 cm tall. Can be reddish. Leaves deeply 3-5-lobed. White 5-petalled flowers up to 6 mm across in spring. On thin dry soils in grasslands, dunes, rocky and urban habitats. Mostly in lowlands.

Plant in April

Leaves in March

Creeping buttercup *Ranunculus repens*

The **yellow, 5-petalled flowers** are similar to those of Meadow Buttercup (page 202) but their stalks are angled and grooved (tubular in Meadow B.) and the **leaves at the base of the plant** have **3 main lobes** with the **end lobe stalked** (so it is really a leaflet). The **side lobes can also be stalked** (again, making them leaflets). All **three lobes/leaflets have their edges cut further into smaller lobes or large teeth**, with a **pale blotch** at each 'cut-in' point (see photo). Some plants, as in the lower-right photo, have more elongated leaves with five lobes/leaflets, four of them in two pairs along a central leaf stalk and the fifth one at the end: these leaves fit better into the previous section of this book (section 3a: lobes/leaflets in rows along left and right sides of leaf) and are indeed mentioned there (page 151). The leaves of Creeping Buttercup are very similar to those of Bulbous Buttercup (page 201), but differences between the two species are: (1) Creeping Buttercup has **shoots that creep just above the ground** (runners), by which means plants can form extensive carpets or patches; Bulbous Buttercup lacks runners and grows as scattered single tufted plants; (2) the flower of Creeping Buttercup has the **sepals pointing out and up** (sepals are down-turned in Bulbous Buttercup). Creeping Buttercup grows mainly on **moist to wet, neutral, fertile soils**, in **woods** (esp. W7), **hedgerows, grasslands** (**unimproved and improved**, esp. MG6-10), **mires and fens** (M22-23 & M27-28, & OV26), **periodically inundated grassy places** (OV28), **gardens** (where it can spread with its runners and become a serious weed) and on **disturbed or waste ground**. It can grow well where there is **nutrient-enrichment** from sources such as runoff from chemically-treated farmland.

Flower in May

Leaves in May

Leaf with 5 main divisions

Bulbous Buttercup *Ranunculus bulbosus*

The lower leaves are like those of Creeping Buttercup (page 200): hairy, with **3 main lobes** and **at least the end lobe stalked**. It differs from Creeping Buttercup in that (1) it has **no above-ground creeping shoots** (runners) but just scattered individual plants, (2) it has a **bulb-like base**, and (3) its flowers have **down-turned sepals** (just below the petals - see photo). The flower stalks are quite short (and straight): mostly 15-25 cm; the plant does not grow as tall as Meadow Buttercup. It grows on **dry to moist, neutral to calcareous soils**, mostly in **unimproved grassland** (MG3 & 5). It is a good **indicator species** for separating **unimproved grasslands** from agriculturally-improved ones.

Flower in May

Leaves in May

Plant with narrower leaf lobes (June); note the spiky, green fruiting head at upper left.

Hairy Buttercup *R. sardous* is similar but is an annual, **lacks a bulbous base** and the **hairs on the leaf stalk stick out** to some degree (closely appressed in Bulbous B.). Mainly in **grassland on disturbed neutral soils** in lowlands of Britain; commonest in S and E.

Meadow buttercup *Ranunculus acris*

The leaves are hairy and **deeply cut into 3-7 narrow, pointed lobes** (lower lobes smallest) that are **quite deeply lobed themselves**. **The lobes are not stalked**, unlike those of Creeping and Bulbous Buttercups. The leaf stalk has a groove running down its upper surface. Upper (stem) leaves are unstalked. The **yellow, 5-petalled flowers** are on branched stems. Their **sepals** (small, greenish and petal-like, below the petals) are **not down-turned**. Flower stalks are tubular - not angled and grooved as in Creeping & Bulbous Buttercups. Grows on **dry to damp acid to basic soils** in **unimproved grasslands**, **hedgerows**, **rush & sedge mires** and **woods**, and on **cliffs**. **Unpalatable** to grazing animals such as sheep, cattle and deer.

Meadow & Wood Crane's-bills (pages 203-204) have leaf lobes more even-sized and leaf stalks ungrooved. Lesser Spearwort (page 295) has narrow unlobed leaves and smaller, narrower-petalled flowers.

Globeflower *Trollius europaeus* also has 'starry' leaves and yellow flowers but the **leaves are hairless, rather shiny** and **wider- lobed**, and **flowers are larger and globular**. In **damp, base-enriched upland habitats:** **unimproved grasslands, wetlands** and **cliff ledges** from Wales northwards; rare in Ireland. **Palatable** to large herbivores.

Wood Crane's-bill *Geranium sylvaticum*

A medium height to tall plant with its lower, larger leaves divided into **7 toothed lobes**. It differs from the superficially similar Meadow Crane's-bill (page 204) in its **wider leaf lobes** with less space between them. The leaves also like those of Meadow Buttercup (page 202) but the lobes are more even-sized, the leaf size can be larger (>10 cm across), the **leaf stalk lacks a groove** along its upper surface, and the **flowers (early summer) are purple** (not yellow). Grows on **dry to moist, neutral soils** in N Britain, in lush, **lightly grazed or ungrazed** vegetation in **woods** (esp. W11), **unimproved grasslands** (esp. MG1 & 3 & U4), **fens** (esp. M23/27) and on upland **cliff ledges** (U17).

Plant in flower in June

Meadow Crane's-bill *Geranium pratense*

A robust plant like Wood Crane's-bill (page 203) but the **leaf**, which is up to 10 cm across, **is more deeply cut** into **5-7 narrower lobes** leaving wider spaces between the lobes. Leaves are like those of Meadow Buttercup (page 202) and Musk Mallow (page 211) but the **leaf stalk lacks a groove along its upper surface**, and the **flowers (midsummer) are purple** (yellow in Meadow B.; pale pink in Musk M.). The leaf is larger, and the whole plant taller, than in Cut-leaved Cranes-bill (page 207). Grows in low-altitude **unimproved, ungrazed or lightly grazed neutral grasslands**, especially MG1 by roadsides.

Plant in flower in July

Bloody Crane's-bill *Geranium sanguineum*

A rather **dense, bushy** plant with **hairy, red-tinged stems**, stalked, hairy, **roundish, very deeply lobed, starry-looking leaves** and, in early to late summer, **large pink flowers**. At low altitudes on **dry, basic soil** in **unimproved, ungrazed or lightly grazed grassland** (esp. CG2), **sand dunes** (SD8/9), **limestone pavements** and **steep, rocky slopes** (e.g. coastal MC9). Mostly along coasts of W & N Britain; scarce in Ireland (mainly mid-W) and inland in Britain; **garden escape on disturbed ground** in lowlands (esp. in England & Wales).

Plants in flower in June

Herb-Robert *Geranium robertianum*

A short to medium-height plant with long-stalked, rather hairy, **triangular leaves cut into 5 leaflets, with each leaflet further divided**. Upper stem leaves can be just 3-lobed. Leaves are **scented** when crushed. The **stems and leaf stalks are hairy**. The stems and leaves are **commonly red-tinged**. There are **small, pink, 5-petalled flowers** in spring and summer. Grows mainly on **dry to moist, neutral to basic soil** in **woods** (esp. W8-9 & 12), **scrub** (W21-22) and **hedgerows**. Also on **cliffs** and in **screes**. Rough Chervil leaves (page 177) can look like those of Herb-Robert but are longer, with leaflets in rows along left & right sides of the leaf stalk. Wood Anemone leaves (page 199) can look similar but are less divided, and their stalks are less hairy and less reddish.

Plant in flower in May

Leaves in September

Smaller Crane's-bills *Geranium* species

Dove's-foot Crane's-bill *Geranium molle* is short, with stalked, **round, hairy, dull green leaves** up to 5 cm across, each with **5-9 small, bluntly-pointed lobes**. **Leaf stalks have hairs up to 1.5 mm long**. **Small pink flowers** with five notched petals 3-6 mm long in spring and summer. Mainly on **dry neutral to basic soils** at low altitude in **short grassland, among rocks** and in **disturbed places** such as **roadsides, field margins** and **waste ground**.

Small-flowered Crane's-bill *G. pusillum* is similar but slightly smaller; leaves <4 cm across with **shorter (<0.2 mm) hairs on leaf stalks**; petals <4 mm long. Mainly on **dry neutral soils** in **lowlands**: in short **grassland** and on **disturbed ground**.

Hedgerow Crane's-bill *G. pyrenaicum* is like Dove's-foot Crane's-bill but **taller** (to 60 cm) with **larger leaves** 5-8 cm across and **larger flowers** with petals 7-10 mm long. Mainly on **dryish, neutral to basic soils** in the **lowlands**: in **grassy**, **little-grazed** vegetation (esp. MG1) in **hedgerows** , **verges** and **field margins**. **Introduced**.

Cut-leaved Crane's-bill *G. dissectum* is like Dove's-foot Crane's-bill but **leaves very deeply cut** into 5-7 very narrow lobes; the leaf looks rather like that of Meadow Crane's-bill (page 204) but smaller. Mainly on **dry, more or less neutral soils** in the **lowlands**: in short **grassland** and on **disturbed ground**. **Introduced.**

Shining Crane's-bill *G. lucidum* is like Dove's-foot Crane's-bill but has **hairless, shiny green or red leaves** and **larger flowers** with petals 8-9 mm long. Mainly **on basic soil among rocks and screes at low altitudes**, especially at the **edges of woods** (W8-9), in **scrub** (W21), in **hedgerows** and on or at the bases of **stone walls**.

Round-leaved Crane's-bill *G. rotundifolium*: like Dove's-foot C. but **leaves shallowly lobed** & **petals not notched**. **Dry lowland banks & disturbed places**, esp. S/E half of Engl.

Dove's-foot Crane's-bill

Shining Crane's-bill

Cut-leaved Crane's-bill

Meadow Saxifrage (p.256) has leaves like Dove's-foot Crane's-bill but **less deeply cut** and with **thicker leaf stalks**; **flowers larger and white**. Mainly in **lowland neutral grasslands**.

Similar-looking roundish leaves with lobes or blunt teeth

Several low-grown species have rather small, roundish leaves c. 1-4 cm across, with edges cut into lobes or large, blunt teeth. Most belong in this section of the book, but some (with leaves less deeply cut) either do not or are borderline in this respect. To help with identification their leaves are all brought together here for comparison.

 **Dove's-foot, Small-flowered and** (larger leaves) **Hedgerow Crane's-bills** (page 207). Leaves **hairy**. Dull green. Veins quite prominent on upper surface.	 **Shining Crane's-bill** (page 207). **Not hairy**. Leaf surface shiny, brightish green or reddish.	 **Meadow Saxifrage** (page 256). Leaves **hairy**, like Alt.-leaved Golden Sax. but dull green with 5-9 shallow lobes/round teeth; leaf stalk widens just below leaf.
 **Ground Ivy** (page 222). Leaves **hairy**, dark green and in **opposite** pairs. Veins prominent on upper surface of leaf (not so in Saxifrages).	 **Opposite-leaved Golden Saxifrage** (page 223). Leaves **sparsely hairy**, yellow-green & in **opposite** pairs; bases not heart-shaped.	 **Alternate-leaved Golden Saxifrage** (page 223). Basal leaf **hairy**; like Meadow Saxifrage but yellower green with >9 rounded teeth.

Two common **Crowfoots** *Ranunculus* species. **Not hairy**. Flowers white, 3-12 mm across, 5-petalled. Scattered shoots or dense mats on **very wet soils** (muddy places, flushes etc).

	**Round-leaved Crowfoot** *R. omiophyllus*. **Acid-neutral** soils, esp. **hill flushes** (M35) in W (N to central Scotl.; in Ireland only in S.		**Ivy-leaved Crowfoot** *R. hederaceus*. Various soils: mainly **neutral** & including nitrogen-rich. Widespread.

Leaves **round** with top of leaf stalk ending in middle of underside of leaf (see page 289).

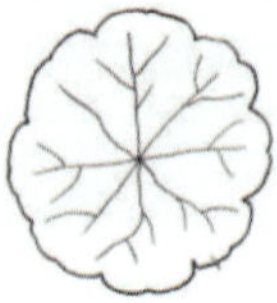	**Marsh Pennywort** (page 289). Leaves **can be slightly hairy**. **Round**. On wet ground.		**Navelwort** (page 289). Leaves **not hairy**. **Round**. Thick, **fleshy** texture. On rocks & tree trunks.

Hemp-agrimony *Eupatorium cannabinum*

A tall plant (up to 1.2 m) whose branched, red-tinged stems bear opposite pairs of unstalked leaves each of which is split into 3 narrow oval, pointed leaflets with toothed edges. Some leaves have 4-5 leaflets. Conspicuous clusters of many small pale pink flowers in summer. On damp to wet, neutral to basic soils, in little-grazed places: fens/mires (esp. M13, 22, 24, 25, 27, S1, S23-26), by rivers and ponds, in wet woods (W2-3 & 5-7), in damp scrub and on damp sea cliffs. Mainly in the lowlands of southern Britain but also along coasts further north.

Plant in July

Sanicle *Sanicula europaea*

Sanicle has leaves deeply divided into 5 lobes with toothed edges. It is easily told from other common species with a similar leaf shape because the leaves are hairless, rather shiny and quite firm-textured. Wood Anemone leaves (page 199) can be hairless but are divided more deeply into three deeply-toothed leaflets. Tiny white flowers grow in small round stalked clusters in summer. On dry to moist, basic soils in woods (esp. ancient W8-9 & W12).

Flowers and leaves in June

Lady's-mantles *Alchemilla* species

There are several species of Lady's-mantle in Britain and Ireland. Our commonest species are small to medium-sized herbs with small to quite large, **long-stalked, roundish, greyish-green leaves** with **7-11 toothed lobes**, folded along the veins so they look **'pleated'**. **Flowers are small and yellow** and grow in **clusters**. Lady's-mantles grow mainly on **dry to moist, neutral to basic soils**, in **unimproved grasslands**, open **woods** and on **rocky banks**.

Leaves of *Alchemilla glabra* in May. Leaf size variable - smaller in short, grazed vegetation.

IDENTIFICATION OF OUR THREE COMMONEST LADY'S-MANTLES:

Species	Upper leaf surface	Leaf stalk
A. filicaulis	hairy	with spreading hairs
A. xanthochlora	hairless	with spreading hairs
A. glabra	hairless	hairless or with closely appressed hairs

Alpine Lady's-mantle in flower in August

Alpine Lady's-mantle *Alchemilla alpina* is different. The stalked leaves are **starry-looking** with **5-7 leaflets** with **silvery-white undersides**. It grows in the **Highlands and Lake District**, on **thin, dry to moist, acid to basic soils**, in **short, moderately to heavily grazed grassland** (CG11-12, U4-5, U7) and ***Vaccinium-Empetrum*** **heaths** (H18-20). Also in **mountain moss heaths** (U10) and **snowbeds** (U14), on **cliffs** and among **rocks by rivers and streams**. Commonest at **moderate to high altitudes**, in places with a **harsh climate**. **Not very palatable** to sheep and deer.

Mallows *Malva* species

Short to tall plants with hairy stems, and stalked, hairy leaves cut into lobes (5 on most leaves) with smaller marginal teeth. Stipules - like tiny oblong leaves - present at bases of leaf stalks. Flowers pink, 5-petalled. On dry, mainly neutral soils in grasslands, hedgerows and waste ground, especially in the lowlands. Not commonly grazed by sheep, cattle etc.

Musk Mallow *M. moschata* has upright shoots (to 80 cm); leaves up to 8 cm across, cut deeply into thin lobes. Stipules distinguish leaf from that of Meadow Buttercup (page 202). Groove along upper surface of leaf stalk distinguishes leaf from that of Crane's-bills (pages 203-207). Flowers pale pink, 3-6 cm across. Native in England, Wales & southernmost Scotland; introduced elsewhere.

Common Mallow *M. sylvestris* is tall and upright (to 1 m) or rather creeping. Leaves are up to 12 cm across, with broad lobes, long stalks and commonly a dark spot at the base of the leaf blade. Flowers are 2.5-4 cm across. Introduced.

Dwarf Mallow *M. neglecta*: like Common Mallow but low, creeping, leaves <7 cm across and only slightly lobed; flowers paler pink, <2.5 cm across. Introduced.

Common Mallow

Dwarf Mallow

Cloudberry *Rubus chamaemorus*

A short plant, mostly <20 cm tall, with roundish, lobed and toothed leaves up to 8 cm across with veins in grooves giving light & shade pattern across leaf. On acid, peaty soils in damp heaths (H21-22) and bogs (M19) in the uplands of Scotland & N England (very rare in Wales & N Ireland). White, 5-petalled flowers 2-3 cm across, like those of Bramble (page 73); by autumn becoming edible red berries.

Young leaf (May)

Old dead leaf

Fruit (July)

Section 3c

PLANTS WITH UNDIVIDED LEAVES GROUPED IN WHORLS OF THREE OR MORE

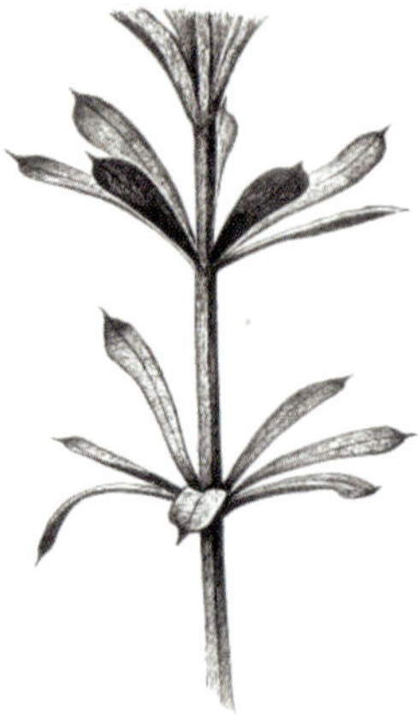

A whorl is a group of three or more leaves growing from the same point along the stem. A whorl of four leaves forms a cross. A whorl of five or more leaves looks star-shaped but is quite different from a single leaf with five or more leaflets radiating out from a central point (as in the previous section of this book).

This group begins with the **Bedstraws** (pages 213-217), which have small, narrow leaves, as in the drawing above, and tiny white or yellow flowers. Some such as Goosegrass (page 217) are scrambling or climbing plants whose stems are too weak to grow tall without the support of other plants. Others are more free-standing but shorter. The Bedstraws are ordered according to the typical number of leaves per whorl: starting with the fewest and ending with the most. A summary of their main features is given on page 213.

Next are three species with larger oval leaves 5-15 cm long. Two of these - **Himalayan Balsam** and **Yellow Loosestrife** - are tall plants that can have some leaves in opposite pairs (or even singly in the case of some upper leaves of Himalayan Balsam). The third species - **Chickweed-wintergreen** - is a short plant with a single whorl of 3-6 leaves that can be of varying size within the whorl.

Finally come the **Horsetails**, which are primitive, unusual-looking plants related to ferns. They have vertical, grooved, green stems and, at intervals along these stems, whorls of thinner green branches. They don't have 'proper' leaves: their equivalent of a leaf is just an inner whorl of tiny scales around the stem, where each group of branches leaves the stem.

Great Horsetail

Other species with whorled leaves: **Juniper**, **Bell Heather** and **Cross-leaved Heath** have whorled leaves but their stems are woody. Juniper is a conifer (see page 49); the other two are dwarf shrubs (see pages 50-71). **Crowberry** (another woody dwarf shrub - see page 67) can have some leaves in whorls. **Purple Loosestrife** (page 252) is a mainly opposite-leaved species that commonly has some leaves in whorls of 3.

Bedstraws – summary table of main identification features

* Note: some (esp. young) shoots can have <6 leaves per whorl	**Leaves per whorl** *Max. leaf length*	**Hairs on leaf margins**	**Hairs on stems**	**Other leaf features**	**Flower colour**
Northern Bedstraw *Galium boreale*	4 *40 mm*	Tiny hairs pointing forwards	Can have tiny projecting hairs	Blunt; 3 parallel veins (1 vein only in leaves of other species); smooth	White
Crosswort *Cruciata laevipes*	4 *25 mm*	Long soft hairs point outwards	Long soft hairs point outwards	Bluntly pointed; softly hairy	Yellow
Squinancy-wort *Asperula cynanchica*	4 *10 mm*	No hairs	No hairs	Pointed; narrow; of unequal length within whorls	White or pink
Field Madder *Sherardia arvensis*	4-6 *12 mm*	V. short hairs pointing forwards	Short hairs point out or backwards	Oval to linear; tapers from middle to fairly sharply pointed tip	Pale mauve -lilac
Marsh Bedstraw *G. palustre*	4-6 *35 mm*	V. short hairs pointing outwards	V. short hairs pointing outwards	Blunt; widest beyond middle; feels slightly rough	White
Heath Bedstraw *G. saxatile*	4-8 *10 mm*	Short hairs pointing forwards	No hairs	Pointed; widest beyond middle; Feels smooth	White
Limestone Bedstraw *G. sterneri*	6-8 * *14 mm*	Short hairs pointing backwards	No hairs	Sharply pointed; widest beyond middle; roughish	White
Fen Bedstraw *G. uliginosum*	6-8 * *10 mm*	Short curved hairs point backwards	Short hairs point backwards	Sharply pointed; widest beyond middle; leaf/stem rough textured	White
Hedge Bedstraw *G. album*	6-8 * *25 mm*	V. short hairs pointing forwards	Stem smooth, or some very short hairs	Sharply pointed; leaf looks & feels smooth	White
Goosegrass / Cleavers *G. aparine*	6-9 * *50 mm*	Short & curved & pointing outwards	Short & curved & pointing backwards	Sharply pointed; widest beyond middle; leaf/stem rough textured	White
Woodruff *G. odoratum*	6-9 * *40 mm*	Short hairs pointing forwards	No hairs	Pointed; looks & feels smooth	White
Lady's Bedstraw *G. verum*	8-12 * *25 mm*	No hairs	No hairs	Pointed; narrow; dark green; feels smooth	Yellow

See also Wild Madder (a large south-western evergreen 'bedstraw') on page 218.

BEDSTRAWS WITH LEAVES IN WHORLS OF 4 (like crosses):

Northern Bedstraw *Galium boreale* has oval, blunt, hairless leaves, each with 3 parallel veins. Small white flowers in summer. Grows on dry to moist, neutral to basic soils in short upland grasslands & among rocks; mainly in Scotland.

Crosswort *Cruciata laevipes* has thick hairy stems with oval, hairy leaves, and small yellow flowers by leaf bases. On dry to moist, neutral soils in ungrazed or lightly grazed grasslands, hedgerows and woodland edges. Common in the lowlands.

Squinancywort *Asperula cynanchica* is short and hairless, with narrow leaves of unequal length within a whorl, and small white or pale pink flowers. On dry calcareous soils in grazed, unimproved calcareous grasslands (esp. CG2) in lowlands N to Cumbria and Yorkshire, and in W Ireland (Galway to Kerry).

BEDSTRAWS WITH LEAVES IN WHORLS OF 4-6:

Marsh Bedstraw *Galium palustre* scrambles up to 70 cm. Narrow bluntish leaves broadest above the middle. Small white flowers. Stems and leaves feel slightly rough. On moist to wet, neutral soils, in short to tall lightly to moderately grazed mires (M22-28), and in woods (W1-7).

Marsh Bedstraw (June)

Field Madder *Sherardia arvensis* creeps and scrambles up to 40 cm. Narrow oval to linear leaves taper from middle to rather sharply pointed tip. Pale mauve/lilac flowers. Roughish stem. On dry soils in lowlands: arable fields, waste ground, dunes, grassland & by paths/tracks. Widespread (but scarce in N & W Scotland + W Ireland).

BEDSTRAWS WITH LEAVES IN WHORLS OF 4-8:

Heath Bedstraw *Galium saxatile* has **smooth, hairless, low grown stems** with **short leaves** 5-10 mm long, **widest above the middle** and with tiny **forward-pointing hairs on their edges**. Small white flowers in summer. On **dry to moist acid soils** in **short, grazed grasslands** (U1-7) & **heaths** (H4 & H8-22), & in **bracken** (U20) & **woods** (mainly W11/17-19). **Unpalatable** to grazing animals such as sheep and deer. Common in short, **heavily grazed** acid grasslands.

Heath Bedstraw in short acid grassland in June

BEDSTRAWS WITH LEAVES IN WHORLS OF 6-8:

Note: young or poorly developed shoots can have <6 leaves per whorl

Limestone Bedstraw *Galium sterneri* is small and low grown, with hairless stems, small leaves and white flowers. It looks like Heath Bedstraw (page 215) but has **backward-pointing prickles on the leaf margins**. It is uncommon, growing in **short unimproved grasslands** (CG9-11) on **basic** soils in uplands.

Fen Bedstraw *Galium uliginosum* is another small-leaved, white-flowered species like Heath, Marsh and Limestone Bedstraws (pages 214, 215and this page), but with **backward-pointing prickles on stems and leaf margins** giving it a **rough texture**: it can look like a small Goosegrass (page 217). **Leaves have sharp points** and are neatly and regularly arranged within each whorl. Like Marsh Bedstraw it climbs among taller plants in **mires/fens** (M22-28) on **wet neutral soils**, but short plants can grow in damp grassland (e.g. CG10b/c).

Hedge Bedstraw *Galium album* is **large**, with **thick, smooth, branching stems** up to 1 m tall, **leaves 8-25 mm long** (smooth-looking but rough-edged with tiny forward-pointing prickles along their margins) and white flowers. Scrambling plants can look like Goosegrass but with smooth stems. Grows on **dry neutral to basic soils** in **unimproved grasslands** (e.g. CG2, MG1 & MG5), **hedges, woodland margins** and **waste ground**, mainly in **southern lowlands**.

Hedge Bedstraw in July

BEDSTRAWS WITH LEAVES IN WHORLS OF 6-12:

Note: young or poorly developed shoots can have <6 leaves per whorl

Goosegrass *Galium aparine* ► has **tall stems scrambling to 2 m** up other plants including hedges and trees. Leaves in whorls of **6-9, widest above the middle**, tapering quite abruptly to a **thin, sharp point** at the extreme tip. **Minute curved hairs** on **stems** and **leaves** make the plant **stick to clothing**. Tiny white flowers in summer. On **fertile, nutrient-rich or eutrophicated soils** in **woods** (W6-9), **scrub** (W21-22), **fens** (M27-28, S26 & 28, & OV26), **hedges, nettle patches** (OV24) and **disturbed** (but **little-grazed and little-trampled**) **ground**. Can be a garden weed. Stinging nettle is a common associate. **Young shoot tips are edible**.

◄ **Woodruff** *G. odoratum* has rather short, erect (non-scrambling) shoots with **oval , rather bright green leaves** that are firmer, larger and broader than those of other common bedstraws and **in whorls of 6-9**. Clusters of small white flowers in spring. On **dryish to moist, basic soils in woods** (mainly **ancient** W8-9).

Lady's Bedstraw *G. verum* ► has **narrow, dark green leaves** in **whorls of 8-12**, and spikes of tiny **yellow** flowers at the shoot tips (in summer). Found mainly in **short**, **grazed**, **unimproved grassland** (esp. CG1-10, MG3/4, U1/4, SD8) **on dry neutral-basic soils**.

One more 'bedstraw' – Wild Madder *Rubia peregrina*

Unmistakeable. **Large; evergreen**; scrambling growth form; tough stems; leaves in whorls of 4–6; rather prickly **backward-pointing teeth on leaves and stems**; **small creamy flowers** in branched clusters; fruits look like **black berries**. In **neutral to calcareous woodland, scrub, hedges** and **rocky places** in SW Britain and S Ireland; very rare in SE England. Commonest near the coast.

Himalayan Balsam *Impatiens glandulifera*

A **tall, hairless** annual, to 2 m, with **reddish stems** with thickenings at intervals, and **oval, toothed, stalked leaves** up to 15 cm long **in whorls of 3** or in **opposite pairs** or (esp. toward shoot tips) just **single**. Leaf arrangement can vary within one plant. **Large pink flowers** in summer and autumn. **Introduced**. Mainly on **damp soils by lowland rivers, streams & ditches**. Also woods and wetlands. **Invasive**, forming large patches. Unusually large for an annual plant.

Flowers and seed pods in October

Opposite leaves in October

Whorl of 3 branches (October)

Plants in August

Stems in October

Chickweed-wintergreen *Lysimachia europaea*

A **short** plant (<20 cm) with **one whorl of 3-6 oval, hairless, more or less unstalked, untoothed leaves of varying sizes** (up to 8 cm long), and **one white flower** about 15 mm across and on a long upright stalk. On **acid soils** in **birch, oak & pine woods** (W11 & W17-18), among **bracken** (U20) and in **heaths** (mainly H12) in **upland areas of N (esp. NE) Britain**.

Yellow Loosestrife *Lysimachia vulgaris*

Tall (to 1.5 m) and **hairy**, with **oval, untoothed, very shortly-stalked leaves** up to 12 cm long and **either in whorls of 3-4 or in opposite pairs**. **Yellow, 5-petalled flowers** up to 15 mm across, in summer. On **wet, more or less neutral soils** in the **lowlands**, mainly in **fens** and **by water**.

Dotted Loosestrife *L. punctata* is similar, but **flowers larger (to 25 mm across) with hairs on edges of petals**. **Introduced.** In various habitats incl. grassland and disturbed ground; mainly in lowlands.

Purple Loosestrife (page 252) can also have leaves in opposite pairs and whorls of three on the same plant (though mostly in opposite pairs, hence its inclusion in section 3d). It differs from Yellow Loosestrife in its square-sectioned stems and pink-purple flowers.

Horsetails *Equisetum* species

Unusual plants with **green, upright, grooved stems** and in most species **groups (whorls) of thin green branches** at intervals up the stems. The **'leaves' are tiny upward-pointing sheaths around the stem at each whorl**. Stems mostly 20-60 cm tall, but Water & Great Horsetails can be >1 m tall. No flowers: just pale brownish, oval, spore-bearing structures at tips of some stems. Related to ferns.

IDENTIFICATION:	**Grooves** on stems	**Sheaths** around stem, at each whorl of branches	**Branches**	**Stem width**	**Hollow centre of stem**
Marsh Horsetail *E. palustre*	4-8 distinct	Loose; grey-green, with 4-8 black teeth edged white	Many; 4-5-angled; hollow	1-3 mm	c.30% of stem diameter
Field Horsetail *E. arvense*	6-18 shallow	Loose; green, with 6-18 black teeth not edged white	Many; 4-angled; solid	3-4 mm	c.30% of stem diameter
Wood Horsetail *E. sylvaticum*	10-18 faint	Loose & long (to 2 cm); green, with 3-6 brown teeth	Many; 3-angled; many thin delicate sub-branches	1-4 mm	25-50% of stem diameter
Water Horsetail *E. fluviatile*	10-30 v. fine, shallow	Tight; green or orange - green +10-30 short, dark, pale-edged teeth	Few or none	2-10 mm	75-80% of stem diameter
Great Horsetail *E. telmateia*	20-40 very faint	Tight; green + 20 -40 long, narrow, brown, pale-edged teeth	Long (to 20 cm); in whorls of 20-40	10-12 mm	40-60% of stem diameter
Rough H. *E. hyemale*	10-30 shallow	Black band at top + bottom, or just short black basal part	None	4-8 mm	>66% of stem diam.
Variegated H. *E. variegatum*	4-10 shallow	Black; white-edged; base can be orange	Few or none; 5-angled; solid	1-2 mm	>33% of stem diam.
Shady H. *E. pratense*	8-20 distinct	6-20; dark + pale edges	Many; 3-angled; drooping tips	1-2 mm	<50% of stem diam.

HABITATS:

Marsh Horsetail *E. palustre*: **wet, neutral to basic soils in flushes & other mires.**

Field Horsetail *E. arvense*: the commonest species; on **dry to moist, neutral soils in grasslands, dunes, flushes, roadsides, gardens and disturbed/waste ground**.

Wood Horsetail *E. sylvaticum*: **moist neutral-acid soils in upland woods, heaths & mires.**

Water Horsetail *E. fluviatile*: mostly in **shallow water**; can form swamps (S10).

Great Horsetail *E. telmateia*: **moist neutral-basic soils; lowland banks, flushes & woods.**

Rough Horsetail *E. hyemale*: **damp soils - flushes, dunes, open woods, by watercourses; widespread; uncommon; commonest in N Britain + N Ireland.**

Variegated Horsetail *E. variegatum*: **damp base-enriched soils - flushes, dune slacks, stony ground; uncommon; mainly in Scotland, N England, Wales & Ireland; v. rare in S + E England.**

Shady Horsetail *E. pratense*: **neutral-calcareous grassland, heath; cliffs; Scotland + N Pennines + N Ireland; uncommon.**

ALSO: Shore Horsetail *E. x litorale*, a hybrid between Field and Water Horsetails. **Intermediate in appearance between the two parent species.** Widespread, especially in **damp disturbed habitats.**

Section 3d

PLANTS WITH UNDIVIDED LEAVES IN OPPOSITE PAIRS

This is a very varied group of herbs. Some are short, some tall, some with large leaves, others small, but all are united in that their leaves are in opposite pairs along the stems and are not divided into lobes or leaflets.

To help narrow things down for identification, the species in this section are ordered as follows:

Leaf margins TOOTHED	Leaves ROUNDISH	Pages 222-224
	Leaves HEART-SHAPED to triangular-oval	Pages 225-229
	Leaves OVAL or RECTANGULAR	Pages 229-237
Leaf margins UNTOOTHED	Leaves SMALL (mostly <4 cm long)	Pages 238-247
	Leaves LARGER (mostly >4 cm long)	Pages 248-252

- Himalayan Balsam and Yellow Loosestrife (pages 219 and 252) are described among the species whose leaves are in whorls of 3 or more, but some of their leaves can be in opposite pairs.
- On very short, young or grazed, plants of Devil's-bit Scabious (page 251) and Red and White Campions (page 250) the opposite leaf arrangement may not be immediately obvious from just the basal leaves alone.
- Common Field-speedwell (page 269) is described in section 3e but some leaves (on non-flowering parts of shoots) are in opposite pairs.
- Although Devil's-bit Scabious (page 251) is in this section some plants have some leaves lobed as in section 3a.
- Field Scabious and Small Scabious are in section 3a (page 165) because most of their leaves are lobed, but their basal leaves can be unlobed as here in section 3d.
- Thyme and Rockrose can look like small, opposite-leaved herbs, but their stems are woody so they are described on pages 68-69 in the Dwarf Shrubs section.
- Butterfly-orchids and Twayblades have just one pair of leaves at the plant base or low down on the stem and are described with other orchids on page 303.
- Sea-purslane has some leaves in opposite pairs but is distinctive in being shrubby, branched (lower stems woody), up to 80 cm tall, with grey-green fleshy narrow oval leaves up to 5 cm long, and growing on saltmarshes: see page 294.

Ground Ivy *Glechoma hederacea*

A **low-grown** plant whose **square-sectioned, creeping or erect stems** bear **dark, dull green, hairy leaves** with **heart-shaped bases**, and **purple flowers**. On **dry to moist, neutral soils** at low altitudes, in **woods** (esp. W8) and **hedgerows**. **Not very palatable** to grazing animals. Can form **ground carpets**, especially where taller species are suppressed by **grazing or disturbance**. Lowest leaves of Garlic Mustard look similar but are brighter green and hairless, and when crushed they smell of garlic (unlike Ground Ivy's 'dry' smell). Opposite-leaved Golden Saxifrage (page 223) has brighter green leaves whose bases are not heart-shaped; it also has yellow flowers and grows in wetter places.

Above: plants in light shade in May. Below: leaves in heavy shade in May

Opposite-leaved Golden Saxifrage

Chrysosplenium oppositifolium

A **low, patch-forming** plant with **bright green, round, bluntly-toothed, sparsely-hairy leaves** and small **yellow-green flowers**. Leaf bases **not heart-shaped**. Grows on **wet, neutral soils** in **woods** (W5 & 7) and in **mires, flushes and springs** (M23, 32 & 37-38). Ground Ivy (page 222) also has small, roundish, bluntly-toothed leaves forming low carpets in shady places, but its leaves are a darker, duller green with heart-shaped bases, its flowers are purple and its habitat is drier.

Flowering plants in woodland in April

Alternate-leaved Golden Saxifrage *C. alternifolium* is similar but its **stem leaves are not in opposite pairs**; its **lower (basal) leaves are on long upright stalks and have heart-shaped bases** (see page 208). On **damp neutral soils**, mainly in **lowland woods & streamsides**.

Meadow Saxifrage *Saxifraga granulata* (pages 208 and 256) has similarly rounded, hairy, stalked leaves with rounded teeth (or short lobes) but again **not in opposite pairs** (mostly at base of plant), and the flowers are larger (to 2 cm across) and **white**. See page 208 for comparison of leaves with the basal leaves of Alternate-leaved Golden Saxifrage.

Eyebright *Euphrasia* species

This plant has **short upright stems** 5-15 cm tall, bearing **small roundish to oval, firm-textured, deeply toothed leaves** about 4-15 mm long. Stems and leaves can be purple-tinged. At the stem tips are short spikes or clusters of **small flowers** that are up to 8 mm across (both vertically and horizontally) but commonly much smaller. The **flowers are mostly white**, with a yellow spot and purplish shading and streaks. Eyebright grows in many types of short or open vegetation including **unimproved, grazed grasslands** (especially on **dry to moist, basic soils**), **open or stony, base-enriched flushes** (M10-12) and various **montane habitats**. There are about 20 species and many more hybrids among these: telling them all apart is a specialist job.

Plants in July. Inset = larger photo of leaves.

Stinging Nettle *Urtica dioica*

This species has **broad oval to heart-shaped**, stalked, **coarsely-toothed leaves** that are 4–8 cm long and in **opposite pairs** up the upright stem. The stems and leaves are clothed with the familiar long, **stinging hairs**. Small green flowers form hanging catkin-like clusters in spring and summer. The stems are commonly branched and grow up to 1 m or more in height, but can die back a lot in winter. Grows mainly on **moist, neutral soils**, especially where **nitrogen- or phosphate-rich**. Commonly an indication of **eutrophication**: for example where there is runoff from fertilised fields, or where grazing animals urinate regularly. Grows in **woods** (esp. W6–9) and **scrub** (W21–23), among **brambles** (W24) and **bracken** (W25), and forming **weedy patches** (OV24–25) **in fields** or on **disturbed** (but **little-grazed and little trampled**) **ground**. Commonly grows with Goosegrass (page 217). **Edible**: young leaves can be made into nettle tea or nettle soup, or cooked and used as you would spinach.

Plants in May

Small Nettle *U. urens* ▸ has **shorter** (mostly <4 cm) **blunter leaves with relatively longer teeth**. On **dry, neutral soils** in **disturbed places**, especially **arable field margins** & **waste ground**. Commonest in eastern lowlands. **Not native.**

Enchanter's Nightshade *Circaea lutetiana*

This species has erect stems with **stalked, triangular, hairless or only sparsely hairy leaves** with **small teeth** along their edges. The stems are up to 60 cm tall and end in long, rather open, airy spikes of **small white flowers** in summer. Grows on **dry to moist, neutral to basic soils** in **woods** (mainly W8-9 & 12-13). Commonest in **ancient woods**. Grows in upland and lowland areas.

Leaves in July

Flowers in July

◄ **Upland Enchanter's Nightshade** *C. x intermedia* (photo) and **Alpine Enchanter's Nightshade** *C. alpina* are **shorter** (mostly <20 cm) and their **leaves have heart-shaped bases and relatively larger teeth**. They grow mainly in **moist or rocky woods in the N and W**, but Alpine Enchanter's Nightshade is scarce. Check for tiny hairs on leaf stalks: these are present on both upper and lower sides of the leaf stalk in Enchanter's Nightshade, on the upper side only in Intermediate E. N. and are absent from the (hairless) leaf stalks of Alpine E. N.

Wood Stitchwort *Stellaria nemorum* (page 248) can look similar in its leaf shape and opposite leaf arrangement, and can grow in similar habitats, but its leaves are untoothed and its flowers much larger, like those of Greater Stitchwort (page 248).

Hedge Woundwort *Stachys sylvatica*

The **large, stalked, heart-shaped, soft-textured, hairy leaves** with **pointed tips** are like those of White Dead-nettle (page 228), but their rather **sharply-pointed teeth** have straight (not curved) edges. Leaves have a distinctive 'dry' **smell**. In summer there are tall spikes of **dull purple flowers**. The **stems are square-sectioned**. Grows on **dry to moist, neutral to basic soils** in **woods** (mainly W8-9) and **hedgerows**. Mainly in places that are **not heavily grazed**.

Plants in June

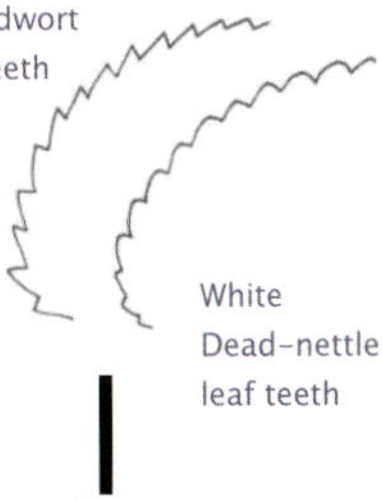

Wood Sage *Teucrium scorodonia*

The **stalked, oval, bluntly pointed leaves are grey-tinged**, with **abruptly tapered or heart-shaped bases**, a **wrinkled texture** and **regularly-spaced blunt teeth** around the edges. They **smell of sage**. The stems are **square-sectioned, branched, quite stiff,** and end in **spikes of small yellow-green flowers**. Mostly on **thin, dry, acid to neutral soils** in **woods** (esp. W8-11 & 15-17), **scrub** (W21-23), among **bracken** (W25), along **hedgerows** and on **cliffs**. Also in some **grasslands** (e.g. MG1-2 & U1), **dry heaths** and **sand dunes**.

Leaves in May

Plant in flower in July

White Dead-nettle *Lamium album*

Rather short (20–40 cm) with **square-sectioned stems and hairy, heart-shaped, toothed, opposite leaves** 4–8 cm long. Leaves resemble those of: (1) Hedge Woundwort (page 227), but with a weaker smell and **teeth with curved, not straight edges** (see drawing on page 227); (2) Stinging Nettle (page 225), but **less coarsely/ sharply-toothed** and **not stinging**; (3) Red Dead-nettle (page 229), but bigger; (4) Yellow Archangel (see below). Upper leaves more oval-triangular. When the **white flowers** are present – at leaf bases, from spring to autumn – identification is easier. On **dryish, neutral soils** at low altitudes, in **lush, little-grazed grassy vegetation** (esp. MG1) along **hedgerows, field margins** and the **edges of woods**. Also on **disturbed** (but **little-grazed and little-trampled**) **or waste ground**. **Not native.**

Yellow Archangel *Lamiastrum galeobdolon*

Like White Dead-nettle but with **low, creeping, carpet-forming stems as well as upright ones. Yellow flowers** in spring. Leaves grow on creeping and upright stems. Leaves are moderately or sparsely hairy, with a less intricate network of veins on their uppersides than in White Dead-nettle. Upper leaves oval and on short stalks; lower leaves more heart-shaped and on longer stalks. Upper leaves with relatively few (<10) teeth on each side (generally >10 in White Dead-nettle). On **moist, neutral to basic soils**, in **woods** and **hedgerows**. Mainly in lowlands. **Native** (subspecies *montanum*) from Cumbria & Yorkshire southwards & in E Ireland. Plants with **big whitish blotches on lower leaves** (ssp. *argentatum* – perhaps a cultivated form of ssp. *montanum*) more widespread as an **invasive introduction**.

White Dead-nettle (April)

Yellow Archangel (April)

Red Dead-nettle *Lamium purpureum*

Mostly <30 cm tall, with **reddish, square-sectioned stems**, **hairy, stalked, blunt-tipped, regularly toothed oval to heart-shaped leaves** 1-5 cm long (upper ones commonly **tinged purple**) and **pink flowers** (spring to autumn). On **dry, neutral soils**, on **roadsides, arable field margins** and **waste ground**: places where vegetation is kept open by **disturbance**. **Less tolerant of trampling** than some other species of disturbed habitats. **Not native**.

Three related non-native species in similar habitats, but less common than Red D-n.:

Henbit Dead-nettle *L. amplexicaule* is **less hairy** and has **leaves with fewer, larger teeth**; **upper leaves unstalked**; **upper leaf-like bracts (adjacent to flowers) clasping the stem**.

Northern Dead-nettle *L. confertum* is like Henbit D-n but larger; **upper leaf-like bracts not clasping the stem**; mainly on Scottish and E Irish coasts + Isle of Man.

Cut-leaved Dead-nettle *L. hybridum* has **leaves stalked**, with **deeply/irregularly toothed edges**.

Hemp-nettles *Galeopsis* species

Medium-height plants (to nearly 1 m) whose **stems have downward-pointing hairs** and are **slightly thicker just below each pair of leaves**. Stems can be branched. **Leaves stalked and oval**, up to 10 cm long, **tapering to both base and tip**; hairy, with **pointed teeth along their edges**. **Flowers pale pink or white**, to 2 cm long; some purple patterning on lower part (the 'lower lip', the tip of which is notched in **Bifid Hemp-nettle** *G. bifida* but not in **Common Hemp-nettle** *G. tetrahit*). Both species grow mainly on **moist to dryish, neutral soils in lowlands**: at the **edges of woods**, in **rush/herb mires**, by **rivers** and **ditches**, on **disturbed/waste ground** and along **arable field margins** (this last habitat being the main one of Bifid Hemp-nettle). **Large-flowered Hemp-nettle** *G. speciosa* has **larger pale yellow flowers** (+ some purple on lower part) **up to 3.5 cm long**; in similar habitats; **not native**.

Red Dead-nettle (April)

Bifid Hemp-nettle (July)

Mints, Marsh Woundwort & Betony

Water mint *Mentha aquatica* is hairy (or, less commonly, hairless), with square-sectioned stems up to 60 cm tall. The leaves smell strongly of mint and are oval, pointed, toothed and short-stalked. Stems and leaves commonly purple-tinged. ink flowers grow in blunt spikes at stem/branch tips and in small clusters among leaves. Mainly on wet, neutral soils in mires/fens/swamps & wet woods. **Corn Mint** *M. arvensis*: similar but no purple tinge; leaves blunt; flowers only among leaves (not at stem/branch tips); in dry to damp grasslands & field/wood margins.

Marsh Woundwort *Stachys palustris* is Mint-like but taller (to 1 m) with longer, narrow oval, toothed and distinctly pointed leaves and no smell of mint. Pink flowers in long, pointed spikes at stem tips. On wet, neutral soils in little-grazed or little-mown places, but also in slightly drier habitats such as damp grasslands.

Betony *Betonica officinalis* is hairy (but hairs short and not obvious) and has dark green oval-oblong leaves with blunt tips and, along their edges, large rounded teeth. Leaf base slightly heart-shaped. Deep pink flowers in blunt-ended spikes. On dryish, more or less neutral soils, in unimproved, grazed or mown grasslands (esp. MG5 & U4) and hedgebanks, in England & Wales. Rare in Scotland & Ireland.

Dog's Mercury *Mercurialis perennis*

This species has upright shoots with mid-green, **oval, pointed, toothed leaves** with **short hairs** and short stalks. There are **small, green catkin-like flower spikes** in spring. It grows on **dry to moist, neutral to basic soils** in **woods** (mainly **ancient** W8-9 & W12-13) where it can form extensive **ground carpets** in spring and summer. It is **not very palatable** to grazing animals such as sheep, cattle and deer. It is restricted to ancient woodland more in the drier southern lowlands of Britain than in cooler, wetter areas further N and W.

Plants in flower in May

A patch of Dog's Mercury in June

Plants forming a ground carpet in August

Annual Mercury *M. annua* looks similar but is **smaller, hairless, annual**, **not native**, and grows on **dry, disturbed ground, including gardens and urban habitats**, mainly in southern British and SE Irish lowlands.

Wood Stitchwort *Stellaria nemorum* leaves (page 248) can look similar but are **untoothed** and **mostly with heart-shaped bases**; the **flowers are larger and white**.

Great Willowherb *Epilobium hirsutum*

The **tall** (up to 1.5 m), **softly hairy stems** are branched in their upper parts and bear **narrow, unstalked, softly hairy leaves with small teeth**, and distinctive **pink, 4-petalled flowers** 2 cm across. The two sides of the leaf base are partly folded against the stem ('clasping' the stem) and are also turned downwards to run a little distance down the stem ('decurrent'). Great Willowherb grows mainly on **moist or wet, neutral to basic soils in lowlands**, along **ditches, hedges, woodland edges** & **watersides**, in **fens** and on **disturbed ground**. It can form dense patches (OV26).

Great Willowherb in July

Hoary Willowherb *E. parviflorum* is similar in appearance, habitats and distribution but is **smaller** (to 60 cm tall), with **leaf bases not clasping the stem and not decurrent down the stem**. Its **flowers are smaller: <9 mm across**.

Broad-leaved Willowherb *Epilobium montanum*

Leaves are **oval, pointed, toothed, short-stalked** , 4-7 cm long, 1.5-3 cm wide and appearing hairless (but tiny hairs visible on very close inspection). Uppermost leaves can be single (not in opposite pairs). Stems mostly 5-40 cm tall and separable from the three next species in being round in section with no raised lines. **Small pale pink 4-petalled flowers** 6-9 mm across in summer. Mainly on **dry to moist, neutral soils**, in **woods** (esp. W7-11), **hedgerows**, **waste ground** and **gardens**. Commonest where vegetation is patchy (as for example in artificially **disturbed** habitats) so competition from other plants is low. Mostly in places that are **not heavily grazed**.

Leaves in June

Flowers in June

American Willowherb *E. ciliatum* resembles Broad-leaved Willowherb but has **2 or 4 raised lines running down the otherwise roundish-sectioned stem**, and the **upper part of the stem has spreading hairs**. **Introduced**. Mainly on artificially **disturbed ground** (waste ground, etc).

Short-fruited Willowherb *E. obscurum* resembles Broad-leaved Willowherb but has **2-4 raised lines running down the otherwise round-sectioned stem**, and its **leaves are unstalked**. On a wide range of soils but commonest on **damp ground** in **various wetland habitats** and at woodland edges; also on **disturbed ground**.

Square-stalked Willowherb *E. tetragonum* resembles Broad-leaved Willowherb but has **distinctly square-sectioned stems** and its **leaves are unstalked**. In England (except Cumbria), Wales and (rare and introduced) Ireland. On **damp or dry soils** in **wetland** and **woodland edge** habitats and on **disturbed ground**.

Marsh Willowherb *E. palustre* (page 243) is like a slender Broad-leaved Willowherb with **leaves narrower** (up to 1 cm wide) and **untoothed or with only tiny, well-spaced teeth**. Grows on **damp or wet, neutral to slightly acid soils** in **rush/sedge mires and flushes** (M5, M23-28) and in **wet woods** (W7). Mostly in places that are **not heavily grazed**.

Bugle *Ajuga reptans*

Short: 10-25 cm. **Square-sectioned stems**. **Leaves oval, more or less hairless, rather glossy, and shallowly toothed**. Stem leaves **stalked.** Lower (basal) leaves **long-stalked**, on long **shoots (runners) just above ground** (like basal leaf of Marsh Valerian - page 189 - but can be toothed and is very gradually tapered at base). **Spikes of blue flowers** in spring/summer. On **moist, neutral soils**, mainly in **woods** (esp. W7; also W8-11).

Plants in May

Red Bartsia *Odontites vernus*

This has **slender stems** that are **commonly branched** (branches widely spreading) and is mostly <30 cm tall. The **leaves are small (1-2 cm long), un-stalked and pointed, with a few teeth along their edges**. In summer there are **spikes of small, dull pink-red flowers**.

Grows on **dry to damp, more or less neutral soils**, in **short grassland** and on **bare or stony ground** by **roads**, **arable fields** and **coasts**.

Yellow-rattle *Rhinanthus minor*

A **rather short** plant (mostly <30 cm) whose **upright stems** are branched or unbranched and have many small **dark spots**. The **small (1-2 cm) leaves** are **narrowly rectangular to triangular, widest at the base, have toothed edges and are unstalked**. The **flowers** have **yellow petals protruding from a roundish, inflated, pale green lower part** (calyx) that becomes pale brown in late summer and autumn, after flowering. Grows on **dry to moist, neutral soil**, especially in **unimproved grasslands** such as **hay meadows** (MG3-5); also in **pastures**, **flushes, fens** and **upland cliff ledges**. Semi-parasitic: can obtain some of its nutrients from roots of other plants, thereby limiting growth of grasses and indirectly encouraging an abundance and good diversity of broader-leaved herbs in grasslands.

Leaves in June

Flowers in June

After flowering

Plants in June

Old dead flower head (November)

Germander Speedwell *Veronica chamaedrys*

A short plant (<20 cm) with slender stems. The **oval to triangular leaves** are hairy, widest below the middle, have toothed edges, are in opposite pairs and are mostly **unstalked** (some on short stalks <5 mm long). **Hairs on stems mainly in two lines on opposite sides of stem. Small pale blue flowers** (spring–summer). On **dry to moist, acid to neutral soils** in **woods** (esp. W11), **grassland** (esp. MG1-2 and some MG5 & U4) and **hedgerows**. It is evidently **not very palatable** to grazing animals.

Wood Speedwell *V. montana* has brighter, yellower green **stalked leaves** (stalks up to 15 mm) and **hairs sticking out all around the stems** (not concentrated in two opposite rows). Mainly on **moist, neutral soils** in **woods** (W7-10), especially **ancient** low altitude woods.

Heath Speedwell *V. officinalis* is more **creeping**, with almost unstalked **leaves tapering gradually to their bases**, smaller leaf teeth, and flowers in denser, narrower upright spikes. Hairs stick out all around stem (as in Wood S.). Mainly on **dryish acid soils** in **woods** (esp. W10-11), **unimproved grasslands** (esp. U4) and **heaths** (esp. H10 & H12).

Germander Speedwell (May

Wood Speedwell (May)

Heath Speedwell (summer)

Common Field-speedwell *V. persica* (p. 269) is rather like Wood Speedwell but leaf stalks <7 mm long, flowers single (not in groups), **flowering shoots with non-opposite leaves**, and very different habitat: **dryish, fertile, open cultivated and waste ground**. **Introduced.**

Teasel *Dipsacus fullonum*

A **tall** plant (up to 2 m) with **stiff, prickly, branched stems** bearing **opposite pairs of large, oval, unstalked leaves** whose main **central vein is prickly on the underside**. Leaf margins have few or many teeth. The **bases of each pair of stem leaves are joined together** to form a 'cup' that can hold water. The **flower heads are large (up to 8 cm long), oval and prickly**. The flowers are **purplish-green in summer**; **old brownish heads persist** through winter and are conspicuous at a distance. Grows mainly on **dry, neutral to basic soils** in the **lowlands**, especially in **tall, little-grazed (or ungrazed) or unmown grassland** (esp. MG1) along **roadsides**, along the **edges of fields and woods**, and on artificially **disturbed ground**.

Small Teasel *D. pilosus*: **smaller** (to 1.5 m); **upper leaves stalked + with a pair of leaflets at their bases** and the **flower heads are spherical**. On **damp calcareous soils**; mainly in the lowlands of England.

Low or creeping herbs with small, opposite, untoothed, oval or roundish leaves - Page 1 of 3 (habitats)

These are short plants (mostly <20 cm); some species can look very alike when not in flower. Their habitats are described on this page. Identification features are on page 239. Photos of most species are on page 240.

Four similar species are described elsewhere: Sea-milkwort and Sea Sandwort (coastal, with thick leaves; see page 241) and Thyme and Rockrose (with woody stems, so they are dwarf shrubs; see pages 68-69). Trailing St John's-wort is also included with other St John's-worts (page 244). See also Lesser Twayblade (p. 303).

Fairy Flax (July)

Heath milkwort *Polygala serpyllifolia*	Mainly on **dry or moist, acid soils**, in **short, grazed** upland **heaths** and **grasslands**. **Not very palatable** to grazing animals. (The similar **Common milkwort** *P. vulgaris* grows in **short, grazed, dry, neutral-basic lowland grasslands**.)
Bog Stitchwort *Stellaria uliginosa*	On **moist to wet, neutral to acid soils**, esp. in **rush/sedge/herb mires**.
Common Mouse-ear *Cerastium fontanum*	Mainly in **short, grazed, unimproved or semi-improved grassland** on **dry neutral to basic soils**.
Field Mouse-ear *C. arvense*	Mainly in **grassland and dunes** on **dry neutral to basic soils**.
Trailing St John's-wort *Hypericum humifusum*	On **dry acid soils**, in **heaths**, at **woodland edges** and on **steep banks**; mainly at low altitude.
Fairy Flax *Linum catharticum*	On **dry to moist, neutral to basic soils**, especially in **short, grazed, unimproved grasslands** (CG1-14), **flushes** (M10-12) and on **rock ledges** (U17).
Blinks *Montia fontana*	On **moist to wet, neutral to acid soils**, esp. in **rush/sedge/herb mires** and **damp grasslands**.
Common Water-starwort *Callitriche stagnalis*	In **shallow water** (in **ditches**, edges of **ponds/rivers** and in **temporarily flooded places** including long-lasting **puddles**) and on **wet muddy ground** (such as rutted tracks, esp. on neutral soils).
Thyme-leaved Speedwell *Veronica serpyllifolia*	Mainly on **neutral soils** in **woods, grasslands, hedgerows, roadsides** and along **tracks and paths**.
Common Chickweed *Stellaria media*	Mainly on **nutrient-rich soils** in **disturbed places**: **waste ground, roadsides, field margins, gardens, stony seashores** and **seabird cliffs**.
Lesser Chickweed *Stellaria pallida*	On **dry, sandy/stony soils** in **open/disturbed** habitats and **sand dunes** in **lowlands** of England, Wales, E. Scotland and E. Ireland.
Scarlet Pimpernel *Lysimachia arvensis*	On **dry, neutral soils** in **disturbed** habitats, especially **field margins** in the **lowlands**.
Yellow Pimpernel *Lysimachia nemorum*	On **damp, neutral soils** in **woods** (esp. W7) and **mires**.
Creeping-Jenny *Lysimachia nummularia*	On **damp, neutral soils** in **woods** and **damp grasslands**, and **by streams**, in **lowlands**. Widespread, but **in Scotland it is scarce and introduced**.
New Zealand Willowherb *Epilobium brunnescens*	**Introduced and naturalised**, on **moist, acid to basic, stony/gravelly soils** (flushes, streamsides, steep banks, edges of tracks etc) in the **uplands**.
Bog Pimpernel *Lysimachia tenella*	On **moist to wet, mainly neutral to basic soils**, in a wide range of **mires** (especially **flushes**; also **dune slacks**). Commonest at **low altitudes** in the **west**.

Low/creeping herbs with small, opp. untoothed oval/round lvs.: p. 2 (identification)

Species ordered more or less by leaf shape: narrower leaves at top of table; rounder leaves at bottom.

	Leaf shape	Leaf hairy ?	Leaf colour	Leaf length	Stem	Flower: colour / no. of petals / diameter
Heath milk-wort *Polygala serpyllifolia* **	Narrow oval; pointed; no stalk; upper leaves not opposite	Not hairy	Dark green	5-15 mm	Not hairy	Dark blue / 2 (sepals)/5-6mm long
Bog Stitchwort *Stellaria uliginosa* *	Narrow oval; pointed; no stalk	Not hairy	Greyish green	10-20 mm	Not hairy; square in section	White / 5 (split almost to base) / 6 mm
Common Mouse-ear *Cerastium fontanum*	Narrow oval; pointed or blunt; no stalk	Hairy	Dull greyish green	10-25 mm	Hairy (hairs 0.5-1 mm)	White / 5 (tips notched) / 6-10 mm
Field Mouse-ear *C. arvense*	As above	Hairy	Dull green	10-20 mm	Hairy (hairs 0.2 mm)	As above but 12-20 mm
Trailing St John-'s-wort *Hyperic-um humifusum*	Narrow oval; blunt; unstalked	Not hairy	Mid green	5-10 mm	Not hairy	Yellow / 5 / 10 mm
Fairy Flax *Linum catharticum*	Narrow oval-oblong; blunt; unstalked	Not hairy	Grey-green	5-10 mm	Not hairy; erect; wiry	White / 5 / 4-6 mm
Blinks *Montia fontana*	Narrow oval (widest beyond middle); blunt; tiny (or no) stalk	Not hairy	Bright green	4-15 mm	Not hairy	White / 5 / 2-3 mm
Common Water-starwort *Call-itriche stagnalis*	Oval; blunt; no stalk	Not hairy	Bright green	4-16 mm	Not hairy; flimsy	No petals; just one stamen (white, 2 mm long)
Thyme-leaved Speedwell *Vero-nica serpyllifolia*	Oval; blunt; tiny stalk or no stalk	Not hairy	Dull green	10-20 mm	Not hairy	Pale blue / 4 / 5-10 mm
Common Chickweed *Stellaria media*	Oval; pointed; un-stalked or on v. short stalk	Not hairy	Bright green	5-25 mm	Line of hairs down stem	White / 5 (split almost to base)/ 5-9 mm
Lesser Chick-weed *S. pallida*	Oval; pointed; on short stalk	Not hairy	Bright green	<7 mm	Line of hairs down stem	Petals tiny or absent
Scarlet Pimper-nel *Lysimachia arvensis* *	Oval; pointed; no stalk	Not hairy	Bright green	15-28 mm	Not hairy; square in section	Red / 5 / 10-14 mm
Yellow Pimper-nel *Lysimachia nemorum*	Oval; pointed; on short stalk	Not hairy	Bright green	10-30 mm	Not hairy	Yellow 5 / 5-10 mm
Creeping Jenny *L. nummularia*	Broad oval; blunt; on short stalk	Not hairy	Bright green	15-30 mm	Not hairy	Yellow / 5 / 15-35 mm
New Zealand Willowherb *Epilobium brunnescens*	Round-oval; tiny stalk; can have a few teeth	Not hairy	Green (purpl-ish be-neath)	3-10 mm	2 lines of tiny hairs run down stem	Pale pink / 4 / 3-4 mm; long-stalked
Bog Pimpernel *Lysimachia tenella*	Roundish; on short stalk	Not hairy	Pale green	5 mm	Not hairy	Pale pink / 5 / 10 mm

* In some places these species can grow taller (to 30 cm or so) by scrambling up among larger plants.

** **Common milkwort** *P. vulgaris* is like Heath milkwort but larger, with branched stems (Heath milkwort is unbranched), longer leaves (up to 2 cm) that are all non-opposite, and longer flowers (up to 8 mm).

Low or creeping herbs with small, opposite, untoothed, oval or roundish leaves - Page 3 of 3 (photographs of most of the species described)

Heath Milkwort

Common Mouse-ear

Bog Stitchwort

Common Water-starwort

Thyme-leaved Speedwell

Common Chickweed

Scarlet Pimpernel

Yellow Pimpernel

Bog Pimpernel

Two small but thick-leaved coastal species

These plants are short - mostly less than 15 cm tall - but their neatly arranged opposite, oval, untoothed, hairless leaves are quite thick and stiff for their small size. They are strictly coastal plants.

Sea-milkwort *Lysimachia maritima* has **slender stems** and its **leaves are dull green** (turning yellow in autumn) **and up to 12 mm long**. **Tiny pink flowers** in summer. Common in **saltmarshes**, especially short saltmarsh vegetation.

Sea Sandwort *Honckenya peploides* has **thicker stems** and **larger, thicker, rigid, brighter green leaves up to 18 mm long**. **Tiny greenish flowers** in summer. Common on **sandy and stony ground along the high tide line**.

Procumbent Pearlwort *Sagina procumbens*

Forms **low, moss-like patches** 1-3 cm tall, with **creeping** shoots and **very narrow, shiny green leaves** 5-12 mm long, each one ending in a tiny point. **Ppale green flowers** (spring to autumn) up to 5 mm across, with 4 petal-like sepals; petals are white, very tiny and just above/inside sepals, or completely absent. Flowers on stalks 1-3 cm long. **Differs from mosses in having flowers, and in its leaves being in opposite pairs**. **Among paving stones**, on **bare** or **trampled ground**, on **paths**, in **gardens** and **by streams**. **In places where vegetation is only short and patchy**.

Sea-spurries *Spergularia* species

Their leaves are also **linear**, but **slightly wider and thicker-textured**, and **flowers are mauve-pink**.

Greater Sea-spurrey *S. media* and **Lesser Sea-spurrey** *S. marina* grow in **saltmarshes** and are more or less **hairless**; flower of Greater S-s about 7-12 mm across; that of Lesser S-s 6-8 mm. Lesser S-s also common **by salted roads inland**.

Rock Sea-spurrey *S. rupicola*: like Greater S-s but with **densely hairy stems**. Grows in south-western **coastal rocky habitats** such as seacliffs and rocky shores, N to SW Scotland and E to Hampshire.

Glasswort *Salicornia* species

Extraordinary greenish to reddish plants whose **short** (<25 cm), **branched, hairless stems** look **thick** and without leaves, but they actually have opposite pairs of tiny leaves held very close to the stems. The flowers are tiny and even harder to see. We have at least seven species but they are very difficult to tell from each other. Collectively they are common in **saltmarshes**, especially in the barer, muddier parts. **Edible**: can be eaten cooked.

Marsh Willowherb *Epilobium palustre*

Slender, solitary stems up to 30 cm tall, with **leaves narrow, blunt, untoothed, unstalked** and hairless or with sparse hairs. Uppermost leaves can be single (not in opposite pairs). **Small pale pink 4-petalled flowers** 4-6 mm across in summer. Grows on **damp or wet, neutral to slightly acid soils** in **rush/sedge mires and flushes** (M5, M23-28) and in **wet woods** (W7). Mostly in places that are **not heavily grazed**.

St John's-worts
Hypericum species

Short to medium-height plants with upright, branched stems, **unstalked oval or oblong, blunt-tipped leaves** and **yellow 5-petalled flowers** 1-2 cm across. Commonest species described below (two of these also described elsewhere: Trailing SJW on page 238, and Tutsan on page 71).

Slender St John's-wort in July: leaves (above); plant in flower (right)

IDENTIFICATION	Stem section	Stem/leaf hairy?	Leaf length
Slender SJW *H. pulchrum*	round	not hairy	0.5-1.5 cm
Hairy SJW *H. hirsutum*		short hairs	2-5 cm
Trailing SJW *H. humifusum* *	2 raised lines down opposite sides of stem	not hairy	0.5-1 cm
Perforate SJW *H. perforatum*			1-2 cm
Tutsan *H. androsaemum* **			5-10 cm
Imperforate SJW *H. maculatum*	square		1-2 cm
Square-stalked SJW *H. tetrapterum*	square + projecting 'wings' on 4 angles		1-2 cm

* stems trailing, not upright as in the other species

** stems woody; flowers become oval, berry-like fruits (first red, then black).

HABITATS	
Slender SJW *H. pulchrum*	**Dryish, acid-neutral soils**, mainly in or along the edges of **woods** (esp. W10-11).
Hairy SJW *H. hirsutum*	**Dryish, neutral to basic soils**, in **grasslands** and **woodland edges/glades** in lowland Britain.
Trailing SJW *H. humifusum*	**Dryish, acid soils**, in **heaths**, at **woodland edges** and on **steep banks**; mainly at low altitude.
Perforate SJW *H. perforatum*	**Dryish, neutral to basic soils**, in **grasslands**, **hedgerows**, **wood edges** & on **disturbed ground**.
Tutsan *H. androsaemum*	**Dry or moist, acid to basic soils**, in **woods** (esp. in W8-11) in S & W; also on cliffs on W coast; introduced in most of E Britain.
Imperforate SJW *H. maculatum*	**Moist, more or less neutral soils**, in **woods** and on **disturbed ground**.
Square-stalked SJW *H. tetrapterum*	**Damp, neutral or nitrogen-rich soils**, in **wet grasslands** and **rushy or herb-rich mires**.

Common Centaury *Centaurium erythraea*

The slender, branched stems are mostly <25 cm tall but can reach 50 cm. The **leaves are <2 cm long, oval, blunt and unstalked**. The pale pink flowers (mid-late summer) are distinctive. On **dry, neutral to basic soils** at low altitudes, in **unimproved short grassland** (CG1-7, U1 & MC5/8/11), **sand dunes** (SD7-8/13/19), and **barer** or **disturbed ground**.

Above & right: plants in June

Gentians *Gentianella* species

Short, branched or unbranched, hairless plants with **unstalked, untoothed leaves** and **upright blue or purple** (rarely white) **flowers** from midsummer to autumn. **Field Gentian** *G. campestris* has flowers with 4 bluish petals. **Autumn Gentian** (also called **Felwort**) *G. amarella* has 4 or 5 purplish petals. Check 'tube' of pointed oval lobes just below petals: **Field Gentian has 4 overlapping lobes of unequal size** (2 big & 2 small) and **Autumn Gentian has 4-5 non-overlapping lobes of more or less uniform size**. Both species grow mainly in **short, grazed, unimproved dry grassland** and sand dunes: **Field Gentian on slightly basic to slightly acid soils** (esp. in CG10 & U4) & mainly in NW; **Autumn Gentian on basic soils** (esp. in CG1-10 & SD8), mainly in England & Wales and widely but thinly scattered in Scotland & Ireland.

Autumn Gentian in August

Marjoram *Origanum vulgare*

This species has branched, slightly hairy stems up to 80 cm tall, with **oval, short-stalked, sparsely-hairy, untoothed or slightly toothed leaves 1.5-4.5 cm long**. Small, **pale pink flowers in dense, stalked clusters** in mid to late summer. Old stems can remain through winter, leafless but with old brown flower/fruit clusters still intact. On **dry, basic soils** in **unimproved, ungrazed or lightly grazed** habitats: **grasslands** (incl. MG1 & CG3-4), banks along hedgerows and roads and among scrub, and on disturbed ground.

Wild Basil *Clinopodium vulgare*

Hairy, with **stalked, blunt leaves with shallow blunt teeth**, and **pink flowers in unstalked clusters** along stem and short blunt flower cluster at stem tip. On **dry, basic soils** in **unimproved grasslands** (e.g. CG3 & 6-7 & MG1-2), **dunes** (esp. SD6-9), **hedgerows** and **sea cliffs** in lowlands of England, Wales & SE Scotland. Similar species: Marjoram (page 246) is less hairy and more branched, with stalked flower clusters; Woundworts (pages 227 & 230) have longer, more pointed leaves and longer, pointed flower spikes at stem tips; Hemp-nettles (page 229) have longer, more pointed and more sharply toothed leaves; Mints (page 230) have minty smell (and Corn Mint lacks flowers at stem tip); Betony (page 230) has darker, less hairy leaves with distinct large rounded teeth and suddenly narrowed bases; Red dead-nettle (page 229) is shorter with heart-shaped leaves and red-tinged, less hairy stems; Self-heal (page 247) is shorter and less hairy, with leaves untoothed and flowers purple and only at stem tip.

Plant in August

Selfheal *Prunella vulgaris*

Has short, upright, square-sectioned stems bearing stalked **small, oval, sparsely hairy leaves** and short blunt flower spikes as in Bugle (page 234), but leaves are **untoothed and not glossy**, and flowers are **purple** (blue in Bugle). On **dry to moist, neutral to basic soils** in **unimproved short grasslands** (MG3-5, CG1-11 & U4), in **rush mires** (M23), **woodland glades** and along **roadsides, tracks and paths**.

Plants in July

Lesser Stitchwort *Stellaria graminea*

Rather straggling, with **thin wiry stems, small leaves** <4 cm long, and **small white flowers** <18 mm across, each with 5 petals that are so deeply-notched that each petal can look like two very narrow petals. Differs from Greater Stitchwort (page 248) in its thinner stems, smaller, smoother-edged leaves lacking a grey tinge, and smaller flowers that are out in midsummer (spring in Greater Stitchwort). Grows mainly on **dryish to moist, acid to neutral soils, in lightly to moderately grazed grassland** (esp. MG1 and U4).

Plant in July

Greater Stitchwort *Stellaria holostea*

The upright, 4-angled, rather brittle stems bear distinctive **leaves: slightly greyish-green, narrow, unstalked, untoothed, rough-edged, widest at the base and with long fine points**; leaf length 4-8 cm. Orientation of leaf pairs alternates by 90° up/down stem, looking like crosses when seen from above. **White flowers** >2 cm across in spring; five petals with deeply notched tips; developing into round, green fruit capsules in summer. Mainly on **dryish to moist, acid to neutral soils**, in **lightly to moderately grazed woodland** (esp. W10-11) and **hedgerows**.

Greater Stitchwort in flower in May

Marsh Stitchwort *S. palustris* (photo below) looks similar but is **slightly smaller** (leaves <5 cm; flowers <2 cm across) and leaves/stems are noticeably **greyer**. In **base-rich wetlands**. Mainly in lowlands of England and central Ireland. Scarce in lowland Scotland and Wales.

Greater Stitchwort: young fruit capsules in June

Wood Stitchwort *S. nemorum* (photo below) has similar flowers (slightly smaller; 13-18 mm across) but **broader, brighter green leaves** (mainly stalked; upper ones not). Leaves and stems can be hairy. On **moist neutral soils in woods** (W6-9) from Wales to S & E Scotland.

Wood Stitchwort in June

Common Cow-wheat *Melampyrum pratense*

Like Greater Stitchwort (page 248) this is a **hairless, rather short plant** (mostly <30 cm tall) with **upright, 4-angled stems** and **narrow, virtually unstalked, pointed leaves** up to 10 cm long. It differs from Greater Stitchwort in its leaves being wider (to 1 cm), greener (not grey-tinged) and its stems not brittle. It is most obvious when the **small yellow flowers** are present, in small groups at leaf bases. It forms patches on **dry, acid soils** in **woods** (esp. W11/17) and **heaths** (esp. H12, 16, 18 & 21-22); mainly in upland areas.

Ragged-robin *Silene flos-cuculi*

Up to 60 cm tall, and unmistakeable when in flower: it has **pink flowers 3-4 cm across, with very deeply-toothed petals**. It can also be identified by its **hairless, rather dark green leaves** up to 10 cm long: **narrow oblong, with pointed tips**. **Lower leaves are widest beyond the middle** and gradually narrowed toward the base where they have short stalks. Stem leaves are widest at or below the middle, and are unstalked. The **main central vein sticks out prominently on the underside of the leaf**. **Stems have sparse downward-pointing hairs**. Grows on **moist to wet, more or less neutral soils**, in various types of **rush-sedge-herb mire** (esp. M5, 9, 13 & 22-28 and SD15 & 17).

Common Cow-wheat (June)

Ragged Robin (June)

Red Campion *Silene dioica*

This species has rather tall, upright, branched, hairy stems with unmistakeable **flowers that are 18–25 mm across and have 5 pink petals with deeply notched tips**. The leaves are **oval, pointed, untoothed, 4–10 cm long and softly hairy on both sides.** The **lower (basal) leaves are on stalks up to 12 cm long and 'winged'** with a very narrow sliver of leaf blade running from the base of the blade down each side of the leaf stalk. The **leaves growing on the stems are almost unstalked**. Forget-me-not leaves (page 288) are similar in shape and texture but narrower, not in opposite pairs and have very indistinct side veins. Red Campion grows on **dry to moist, neutral to basic soils** in **light shade**: in **woods** (esp. W8–9), along **hedgerows** and on **cliffs**. It appears to be **moderately palatable** to grazing animals.

Plant in May

Lower (basal) leaves in June

White Campion *S. latifolia* is similar but the **flowers are white** and slightly larger, **and the lower leaves have unwinged stalks** up to 5 cm long (up to 12 cm in Red Campion). On **dry, neutral soils** in **hedgerows** and **field margins**, and on **disturbed or waste ground**. Mainly in lowland areas. Unlike Red Campion it is **uncommon in woods**. **Not native**. **Hybrids** between the two species are quite common and have **pale pink flowers**.

Devil's-bit Scabious *Succisa pratensis*

Flowering plants are distinctive with their **round bluish flower heads** on branched stems up to 1 m tall. **Leaves** are mostly at the base of the plant: **stalked, oval, pointed or blunt, untoothed, sparsely hairy and about 5–20 cm long; pale undersides with an intricate network of darker veins** showing a clearer pattern than in the otherwise quite similar leaves of Common Knapweed (page 276; rougher textured; can have some teeth or small lobes) and Red & White Campions (page 250; softer, hairier). On **moist, acid to neutral soils**, in **mires & fens** (M4–6, 9–11, 13, 15–16 & 22–27), **heaths** (mainly M15–16, H5 & H7), **unimproved grasslands** (MG2–3, 5 & 8–9, CG2, 6, 9 & 10–14, & U4–6) and **cliff ledges** (U16–17). **Palatable** to grazing animals but can survive as low, non-flowering plants in short, grazed vegetation.

Flowering plants in September

Basal leaves in October

Close up of leaves from base of plant (October)

Purple Loosestrife *Lythrum salicaria*

Tall (up to 1.2 m), with **4-angled stems** and few or no branches. **Leaves unstalked, oblong, untoothed** and 4–7 cm long, arranged mostly in **opposite pairs** but some can be in **groups (whorls) of 3**, or, on the upper parts of stems, just single. Stems and branches have at least some very short, fine hairs. **Flowers are pink-purple**, 10–15 mm across, in conspicuous **long spikes**. Grows on **wet, neutral to basic soils**, in lush **ungrazed to lightly grazed wetlands** (esp. M9, M13, M22–24, M27–28 and S24–28). Widespread. Mainly at low altitudes. Scarce in N Scotland. Unknown in Orkney/Shetland.

Yellow Loosestrife also has leaves in pairs or whorls of three, but mostly in threes (hence its inclusion in section 3c) but differs in its rounded (not square-sectioned) stems and yellow flowers. Without flowers Purple Loosestrife can resemble Ragged robin (p. 249) and Willowherbs (pp. 232–233), but their stems are not 4-angled except in Square-stalked Willowherb (p. 233), which has toothed leaves.

Plant in July

Wild Clary *Salvia verbenaca* has branched stems up to 80 cm tall, **hairy grey-green leaves with large teeth** and **spikes of small purple flowers**. On **dry soils in open lowland habitats**. Rare in Scotland/Irel.

Wild Clary

Section 3e

PLANTS WITH UNDIVIDED LEAVES NOT IN WHORLS OR OPPOSITE PAIRS

These have leaves NOT divided into lobes or leaflets, NOT grouped into whorls and NOT in opposite pairs. This is a large and varied group within which the species are ordered as follows:

		Start at:
Species whose leaves are HEART-SHAPED *		Page 254
Species whose leaves are NOT HEART-SHAPED **	Side veins BRANCH OUTWARDS off the main central vein	Page 261
	Side veins run PARALLEL to main central vein or appear to be absent	Page 291

* Giant Knotweed has heart-shaped leaf bases but is described on page 262.

** In Harebell (page 287) the roundish basal leaves have clear outward-branching side veins, but the stem leaves can be so narrow that side veins are not apparent.

Black Bryony and Hedge Bindweed should strictly be in this section, but because of their distinctive climbing growth form they are described on page 78 (in section 1d) among the similar-looking (but woody) climbing and scrambling plants.

Sneezewort should be in this section but is described on page 174 alongside the related Yarrow. It has narrow, pointed, dark green leaves edged with numerous regularly-spaced very small sharp teeth, and clusters of white flowers at the tops of the stems, which are up to 60 cm tall.

Common Milkwort, a short plant with small oval leaves and small blue flowers, qualifies for inclusion here but is described on page 238 (in section 3d) alongside Heath Milkwort.

Narrow-leaved Ragwort should be in this section but is on page 168 with other ragworts, in section 3a.

Alternate-leaved Golden Saxifrage, a short plant with roundish leaves and small yellow flowers, belongs here but is described on page 223 (in section 3d) alongside the similar and closely-related Opposite-leaved Golden Saxifrage.

Himalayan Balsam is described on page 218 in section 3c (species with leaves in whorls), but some of its leaves are in opposite pairs (as in section 3d) or borne singly (as here in section 3e).

Greater Knapweed and Wood & Water Avens are described on pages 164 and 190 in section 3a (lobes/leaflets along L & R sides of leaf), but they can have some leaves unlobed - such leaves would strictly fit better in this section.

Nipplewort (page 284), Goosefoots & Oraches (page 265) and Oil-seed Rape (page 268) - all in this section - can have some leaf lobes (in Nipplewort just at bases of some lower leaves): this is a point of similarity with the species in section 3a (lobes/leaflets along L & R sides of leaf).

Common Field-speedwell (page 269) has some leaves in opposite pairs as in section 3d.

Lesser Celandine *Ficaria verna*

A **short** plant, typically growing no more than 10 cm tall. It has **glossy, hairless, heart-shaped, long-stalked leaves** that are 1–4 cm long and more or less untoothed. Some leaves have paler or darker patches on them. The flowers (in spring) are 2–3 cm wide, with **8–12 narrow yellow petals**. Lesser Celandine grows on **moist, neutral soils** in **woods** (W7–9), **scrub** (W21–22), among **bracken** (W25/U20), in **unimproved grassland** (mainly MG1–2) and in **hedgerows**. It can form carpets of leaves in spring, but dies back by autumn.

Plants in April

Marsh-marigold *Caltha palustris*

Short to medium height plants with thick stems and lush growths of **hairless, heart-shaped, long-stalked leaves** up to 10 cm long, with bluntly toothed edges. Leaves are similar to those of Lesser Celandine but larger, toothed, and lacking any paler or darker patches. **Large, buttercup-like yellow flowers** in spring are up to 5 cm across and are 5-petalled. Grows on **moist to wet, neutral soils**: mainly in **mires and fens** (M9, M12-13, M22-28, S24-25, S27 and MG8), **springs** (M23 and M37-38) and **wet woods** (W1-7). The variety *radicans* grows on mountains and is smaller, with flowers <3 cm across.

All photos: plants in April

Garlic Mustard *Alliaria petiolata*

Also called **Jack-by-the-Hedge**. **Tall and hairless**, with **stalked, bright green, triangular to heart-shaped, coarsely-toothed leaves** growing at base of plant and well up stems. **Clusters of small white 4-petalled flowers** distinctive in spring and early summer. Grows on **dry to moist, neutral and basic soils**, along **hedgerows, walls and woodland margins**, and on **waste ground**. Lowest leaves can be roundish and like those of Ground Ivy (page 222) but are brighter green and hairless (not dark, dull green and hairy as in Ground Ivy) and **smell of garlic** when crushed.

Meadow Saxifrage *Saxifraga granulata*

This has **round to heart-shaped, hairy, stalked leaves** mostly at **plant base**, with 5-9 **large rounded teeth** (could also be described as shallow rounded lobes). **White, 5-petalled flowers**, to 2 cm across, on upright stems in spring-summer. Leaves can look like those of Golden Saxifrages (p. 223) and small Crane's-bills (p. 207): all of these are compared on page 208. Mainly in **lowland unimproved neutral grassland** (esp. MG5); also woods (esp. W7-8). Very rare in Ireland.

Coltsfoot *Tussilago farfara*

The **yellow dandelion-like flowers** appear on **thick, scaly stalks** in early spring, **before** the leaves. Later they become round white fluffy seed heads. The leaves are long-stalked, 10-20 cm long and wide, **thick, firm-textured** and **heart-shaped** with irregular edges with **well-spaced tiny dark teeth**. Young leaves have whitish felty down on both sides, soon coming off the upperside but staying on the underside. On **dry or moist, mainly neutral soils**, especially where **disturbance** (**natural** or **artificial**) keeps vegetation **open or short**: **road verges, soily banks, stony or shingly ground, sand dunes** and **waste ground**.

Flowers in April

Leaves in April

Butterbur *Petasites hybridus*

This has **spikes of pale pink flowers** in **early spring, before the leaves emerge**. The **tough, long-stalked, heart-shaped leaves** are like those of Coltsfoot but with **many small green teeth along their edges**. They grow **very large** (50 cm or more across on long, ridged/channelled stalks) but die back in winter. On **moist, fertile, neutral soils** by **lowland rivers, streams** (including W6 **riverside woods**) and **roadsides**.

Flower spike (April, with Ground Elder leaves)

Leaves in July

Four similar **introduced** species. **Giant Butterbur** *P. japonicus* has 12-14 parallel ridges on underside of leaf stalk (7-10 in Butterbur); in similar habitats but less common. **Winter Heliotrope** *P. fragrans*: **more open flower heads**; **smaller leaves** <20 cm wide **persisting through winter**; leaf stalks **smooth** (not ridged/channelled); in similar habitats. **White Butterbur** *P. albus*: **white flowers**; **smaller leaves** <30 cm wide **with dark uppersides, white undersides and jaggy teeth**; leaf stalks **smooth**; **woods** and **disturbed ground** in **lowlands** (esp. E Scotland). **Leopardsbane** *Doronicum pardalianches*: hairy basal leaves <8 cm across with channelled stalks, appearing in early spring; tall branched hairy stems with unstalked oval leaves and **large yellow daisy-like flowers** in spring-summer; on moist, fertile soils in lowland **woods**.

Lesser/Wood Burdock *Arctium minus s.l.*

The **lower leaves are large** (up to 40 cm long), **heart-shaped, longer than wide, with pointed tips** and untoothed or with small, widely-spaced teeth along their edges. They are **thick and tough**, but their surfaces have a rather **soft texture (very short, soft hairs)**. Lower leaves are on **thick, furrowed stalks** up to 30 cm long. **Leaves on stems are smaller and more oval-shaped**, with shorter stalks. Main stems are thick, **branched**, and can grow over 1 m tall: in summer they have **dense, roundish, green-purple flower heads** up to 3 cm long, becoming **brown fruiting heads** in autumn; both flowers and fruits have **bristles whose hooked tips catch on our clothing**. Grows mostly on **dryish neutral soils** in the **lowlands**, in **woods** (especially woodland edges), **hedgerows, field margins** and on **disturbed or waste ground**.

Note: *s.l.* = *sensu lato*. Lesser and Wood Burdocks *A. minus* and *A. nemorosum*, which are very similar to each other, are treated together here as one.

Greater Burdock *A. lappa* has **leaves as wide as they are long, with greyer undersides and solid stalks** (hollow in Lesser Burdock). **Introduced**. Mainly on **disturbed ground** and along **hedgerows** and **riverbanks** in the **lowlands of southern Britain**.

Water-lilies

White Water-lily *Nymphaea alba* has floating leaves up to 30 cm long with veins radiating out from centre, and large white flowers up to 20 cm across. The whole plant is floating. Grows in undisturbed, mainly unpolluted, still or slow-moving, acid, neutral or basic water in lowland and upland areas.

Yellow Water-lily *Nuphar lutea* has floating leaves up to 30 cm long with most veins running out either side of a central main vein, and smaller yellow flowers 4-6 cm across. The plants are rooted into the ground. Grows in still or slow-moving, acid, neutral or basic water in the lowlands. Can tolerate some eutrophication.

Least Water-lily *N. pumila* is like a small Yellow Water-lily with leaves up to 13 cm long and flowers up to 3.5 cm across. In acid to neutral Scottish lochs.

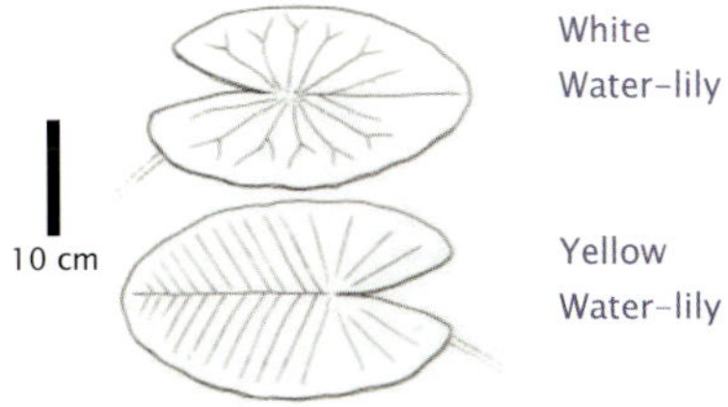

Lords-and-ladies *Arum maculatum*

Also called **Cuckoo-pint**. Unmistakable with its large, long-stalked, hairless, heart-shaped leaves, that can have dark blotches on them, arising in spring from the base of the plant, and the unusual flower (April-May) that has poisonous red berries in late summer. Mainly on moist, neutral soils, in woods and hedges in the lowlands. Native in England, Wales & Ireland. Introduced in Scotland & Isle of Man.

Dark brown flower becomes a spiked cluster of poisonous red berries in late summer

Italian Lords-and-ladies *A. italicum*: similar but leaves arise in winter; all unspotted and with pale midrib. Yellow flowers. In lowland woods, hedges, etc. Native in far S of England; introduced elsewhere.

Violets *Viola* species

Short plants with **heart-shaped, long-stalked leaves** (mostly at plant base) and **purple flowers**. Evidently **not very palatable** to grazing animals. In the following descriptions the number of teeth on each side of the leaf is given (e.g. 13-17T).

Common Dog-violet *V. riviniana* has **mid to dark green leaves (13-17T)** slightly longer than wide, and **mid purple flowers** (spring) with **pale cream 'spur'** sticking out at back of top of flower. On **dry to moist, neutral to basic soils** in **woods** (W8-12), **scrub** (W19 & 21-22), **bracken** (U20), short **grasslands** (CG9-14 & U4-5), **hedgerows** and some **heaths** (H8/10). In short, **grazed vegetation**; also in **less grazed places** where vegetation is not too tall or thick (e.g. on steep rocky banks).

Marsh Violet *V. palustris* has **rounder, blunter, paler, brighter green leaves**, and **paler flowers** (late spring - early summer) and grows mainly on **wet, acid to neutral soils**, in **rush/sedge mires** (M4-8) and **wet woodland** (W4 & 7).

Common Dog-violet in April

Marsh Violet flower (May) and leaves (July)

Early Dog-violet *V. reichenbachiana* is like Common Dog-violet but with a narrower **dark purple 'spur'**; in **woods on basic soils** (W8 & 12) **in southern lowlands**.

Hairy Violet *V. hirta* is like Common Dog-violet but **leaves (16-30T) longer than wide, hairy** and **leaf stalks with sticking-out hairs 0.3-1.0 mm long**. In **grasslands & wood margins on neutral-basic soils in lowlands** (mainly in the south).

Sweet Violet *V. odorata* is like Common Dog-violet but **leaves (20-25T) broad & rounded**, & **leaf stalks with backward-pointing hairs <0.3 mm long**. In **hedges and woodland edges, on neutral to basic soils in lowlands** (mainly in the south).

Heath Dog-violet *V. canina* is like Common Dog-v. but **leaves (10-15T) thicker, narrower than long, with bases only slightly heart-shaped**, and with **blue flowers**; rather uncommon in **heaths, grasslands & dunes** on **acid to basic soils in lowlands.**

Pale Dog-violet *V. lactea* has pale purple flowers but unlike all the above species it has **narrow oval leaves (10-14T) with gradually tapering bases that are not heart-shaped**. Uncommon. **Dry acid lowland heaths** in W Wales, S England and S Ireland.

Pansies *Viola arvensis/tricolor/lutea*

Short plants with **oval, toothed, stalked leaves**, and **flowers purple, yellow, white or cream** (or mixture). Check flower size and stipule shape (stipules are like small leaves at the bases of the leaf stalks - see drawings below).

Field Pansy *V. arvensis* has **small flowers** (8-20 mm from top to bottom) that are **white or cream** with some yellow and purple-blue in centre. **End lobe of stipule is large**, almost like a leaf (smaller in the other two species - see drawings). **Introduced**; on **dry, more or less neutral soils** in **disturbed** places (**arable fields, waste ground** etc) in **lowlands**.

Wild Pansy *V. tricolor* has **larger flowers** (15-25 mm top-bottom) **varying from yellow to purple** as in Mountain Pansy photos below. **End lobe of stipule** (see drawings) is **narrow** in ssp. *tricolor* (**dry acid to basic soils in grassland** & on **disturbed/waste ground** - mainly in **lowlands**), **or wider than side lobes** in ssp. *curtisii* (**coastal grassland** and **sand dunes**).

Mountain Pansy *V. lutea* has **even larger flowers** (20-35 mm top-bottom) **varying from yellow to purple** (see photos); **upper leaf surface has sparse short, stiff, white hairs**; **stipule** (see drawing) **has small narrow end lobe**. On **dry, acid to basic soils** in **short, grazed, upland grassland** (esp. U4 & CG10).

Colour variation in Mountain Pansy flowers

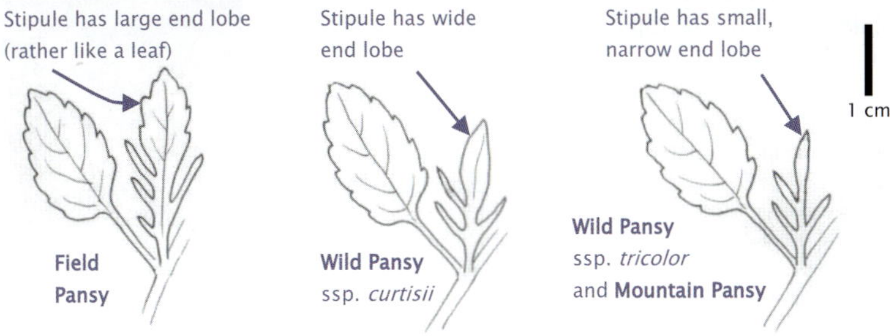

Japanese Knotweed *Reynoutria japonica*

The **branched, arching, reddish or greyish-green stems are up to 2 m tall** and bear **large, stalked, triangular** or nearly heart-shaped, **hairless leaves** up to 16 cm long. **Spikes of small white flowers** in late summer to autumn. **Introduced**. **Invasive** on **disturbed and waste ground**, especially **near human habitation in lowlands**. Prone to spreading along **banks of rivers and streams**.

Giant Knotweed *R. sachalinensis* is similar but **larger**: up to 4 m tall; **leaves up to 30 cm long** with **heart-shaped bases**. **Introduced.** In similar habitats to Japanese Knotw. but less common.

Knotgrasses *Polygonum*

Low and spreading with **short oval leaves with silvery to pinkish sheath-like stipules at their bases. Flowers tiny; white to pale pink**. On open **disturbed/waste ground**.

Knotgrass *P. aviculare* * has **main stem leaves larger** (to 4 cm long) **than outer branch leaves**; native. **Equal-leaved K.** *P. arenastrum* has **leaves all about the same size** (<1.5 cm long) and is **introduced**. Both species are widespread and common. **Cornfield K.** *P. rurivagum* * has **leaves distinctly narrow** (width <17% of length); **introduced** in **arable fields in England/Wales lowlands**; rare in Scotland; not in Ireland. **Ray's K.** *P. oxyspermum* has **bluer/greyer leaves <4 cm long**; on **stony/sandy coastlines**. **Northern K.** *P. boreale* * is like Knotgrass but has **leaves held well clear of stipules, on stalks up to 8 mm long** (leaf stalks <3 mm in other Knotgrass species); native on **Scottish coastal sand/shingle**. (* = Some or many shoots can be upright.)

Docks

Broad-leaved Dock *Rumex obtusifolius*. A hairless plant with tall, stiff, branched stems. Leaves tough, bright green, oval-oblong, up to 25 cm long and 2-3 times as long as wide, with bluntly pointed tips and abruptly tapered (or heart-shaped) bases. Has spikes of small, green to red flowers and fruits. Plants turn brown in autumn. Mainly on neutral soils in disturbed places: waste ground, roadsides, hedgerows, field margins, riversides and heavily grazed or trampled pastures. Mostly where disturbance prevents total vegetation cover. Can be a garden weed.

Wood Dock *R. sanguineus* and **Clustered Dock** *R. conglomeratus* have narrower leaves 4-5 times as long as wide, with more pointed tips and more tapering bases. Wood Dock has straightish stems with branches pointing upwards, and grows on moist neutral soils in lowland woods (W6-8, W10 & W12). Clustered Dock has wavy, zig-zag stems with branches sticking out widely, many small leaves on flower spikes, and grows on damp neutral soils in lowlands - in grasslands, by ditches, rivers & streams, and in periodically flooded places; not in woods.

Curled Dock *R. crispus* has dark or dull green leaves that are narrow (5-6 times as long as wide), taper gradually to their bases and have very wavy edges. In similar disturbed habitats to those of Broad-leaved Dock; also along upper parts of seashores and in sand dunes.

Northern Dock *R. longifolius*: like Curled Dock but fruit blunter-tipped and lacking central swollen part. On disturbed ground in Scotland and N England.

Broad-leaved Dock in June

Broad-leaved Dock (Sept)

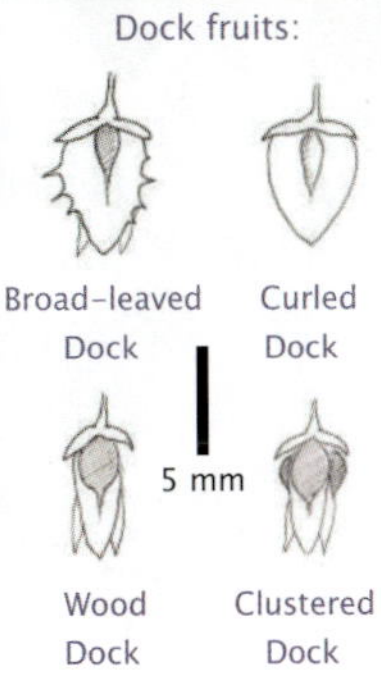

Common Sorrel *Rumex acetosa*

A herb with **long-stalked, oval or oblong mid green leaves**, each of which has **two pointed lobes at its base**, these lobes **pointing backwards** more or less in line with the leaf stalk. The upright stems end in long, narrow, branched spikes of small reddish flowers like those of docks on page 263 (to which Sorrels are related). The **leaves are edible**: they have an acid taste. Common Sorrel grows mostly on **dry to damp, neutral to acid soils,** in **short to rather lush**, **moderately grazed to ungrazed** vegetation: **grasslands** (MG1-10 & U4), **mires** (M22-28), **cliff ledges** (U17, MC3-4, 7-9 & 11-12) and among **ferns** (U18-20).

Common Sorrel leaf in July

Sheep's Sorrel *Rumex acetosella*

Like Common Sorrel but smaller, with **narrower, greyer leaves** whose **basal lobes stick out sideways**. **Leaves are edible** (acid taste). Mostly on **thin, dry acid soil in short, sparse or heavily grazed vegetation** in **well-lit** places such as short, stony **lowland acid grassland** (U1); soils thinner and vegetation more open than in typical Common Sorrel habitat.

Sheep's Sorrel in July

Sea Beet *Beta vulgaris* ssp. *maritima* has **shiny dock-like edible leaves** and **spikes of green flowers**. Forms **branched clumps** on **coastal cliffs** and **strandlines**, especially where soils are **nutrient-rich**. Widespread, but rare in N and E Scotland.

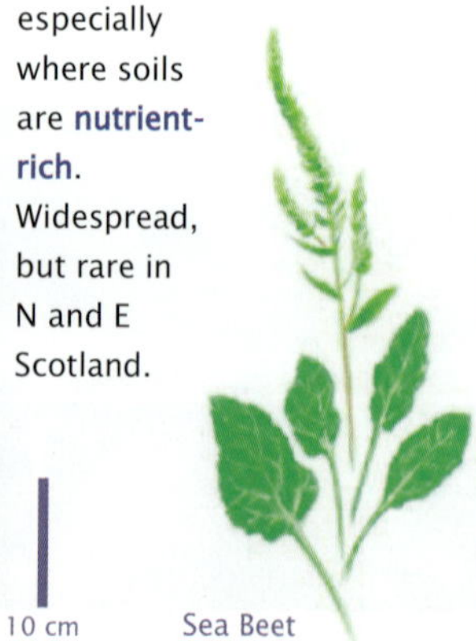

Sea Beet

Goosefoots *Chenopodium* & Oraches *Atriplex*

Weedy-looking dull grey-green or red-tinged plants with branched, upright or low and spreading stems. Most **leaves oval, diamond-shaped or triangular, with some prominent large blunt teeth**, but upper leaves can be narrower and less toothed (and Grass-leaved Orache has all leaves very narrow, with few or no teeth). **Flowers small and dull grey-green or reddish**. Main habitats are **sandy or stony coastal tide-lines**, and **disturbed ground (rubbish dumps, roadsides etc) inland**. Many species: can be difficult to separate from each other but this table summarizes the main features of the commonest species. See also Sea-purslane (page 294).

Species	Stems creeping or upright	Stem colour	Upper surface of leaf	Angle formed by L & R sides of leaf base	Grows inland or only along coasts
Red Goosefoot *Chenopodium rubrum*	Vary from creeping to upright	Reddish	Glossy and hairless	<90° to >90°	Mainly inland
Fat-hen *Chenopodium album*	Upright	Green or red	Matt 'dusted' grey-green finish	<90°	Inland
Common Orache *Atriplex patula*	Vary from creeping to upright	Green or red	Matt 'dusted' grey-green finish	<90°	Inland and coastal (but not tide-line)
Spear-leaved Orache *Atriplex prostrata*	Vary from creeping to upright	Green or red	Matt 'dusted' grey-green finish	Many leaves >90°	Inland and coastal (esp. tide-line)
Babington's Orache *Atriplex glabriuscula*	Creeping	Green or red	Matt 'dusted' grey-green finish	Many leaves >90°	Coastal tide-line
Grass-leaved Orache *Atriplex littoralis*	Upright	Green or red	Matt 'dusted' grey-green finish	<90° (all leaves very narrow)	Coastal tide-line
Frosted Orache *Atriplex laciniata*	Creeping	Yellow or red	Very 'dusted' matt silvery finish	About 90°	Coastal tide-line

L & Upper-R: Spear-leaved Orache

Lower-R: Frosted Orache

Bistorts *Persicaria* species

In these plants the **base of the leaf stalk is enclosed by a sheathing 'stipule'** (see photo) as in Docks (p. 263) and Sorrels (p. 264). The **leaves are oval and in some species have a distinct blackish blotch on the upperside**. The **flowers are small, pale pink and in dense spikes**. The six commonest species are described below.

MAIN IDENTIFICATION FEATURES:	**Tapering of leaf base**	**Dark blotch on leaf upperside?**	**Leaf underside**
Redshank *P. maculosa*	Gradual	Common	Almost hairless
Pale Persicaria *P. lapathifolia*	Gradual	Common	Sparse to dense white woolly hairs
Amphibious Bistort *P. amphibia*	Abrupt (rounded)	Common	Hairless or with very short hairs
Water-pepper *P. hydropiper*	Gradual	Absent	Hairless
Common Bistort *P. bistorta*	Abrupt (round to heart-shaped)	Absent	Almost hairless
Alpine Bistort *P. vivipara*	Gradual	Absent (upper-side dk. green)	Hairless; pale; v. raised central vein

Redshank in October

Sheathing stipule (this example in Redshank)

Common Bistort in June

HABITATS:

Redshank	Damp nutrient-rich soil in grassland d, arable/disturbed land & by water.
Pale Persicaria	Various damp soils in grassland, arable/disturbed land & by water.
Amphibious Bistort	Wide range of wet habitats, from aquatic (in still or slow-moving water) to terrestrial (on very wet, more or less neutral soils).
Water-pepper	Wet mud (neutral to acid) by water, at field edges, along tracks etc.
Common Bistort	Ungrazed to lightly grazed grasslands, hedgerows & steep banks, on neutral to acid soils; rare in Ireland. Also a garden escape.
Alpine Bistort	Grassland & flushes on base-rich soil in mountain areas in N.

Hart's-tongue *Asplenium scolopendrium*

This fern is unmistakeable with its **tufts** of **bright, glossy green, oblong leaves** (fern leaves are more correctly called 'fronds') that are **evergreen** and have **heart-shaped bases, pointed tips** and a **tough, leathery texture**. The fronds are on stalks that are shorter than the fronds themselves, and are stiff and covered with brown scales. As in all ferns there are no flowers. Clusters of tiny spores (reproductive structures) form long, narrow, brown stripes on the undersides of fronds. Hart's-tongue grows on **neutral to basic soils** in **shaded habitats**, especially in **woods** (W8 and to a lesser extent W9 and W12, and W21-22 scrub), among **rocks** (including crevices in limestone pavements), on **walls** and on **steep banks**.

Plant in September

Oil-seed Rape *Brassica napus*

This species is **planted as a crop** in **lowland fields**, but can spread onto **disturbed or waste ground**. Rape fields are **conspicuous in spring** when the plants are in flower. Oil-seed Rape is mostly **hairless**, with **tall** (1 m) **grey-green, branched stems** with clusters of scented, **yellow, 4-petalled flowers** at their tips. The **greyish-green leaves** are **oval to oblong**; the **lower ones have side lobes and long stalks; those up the stems** – more readily seen than the basal leaves – **are unlobed, unstalked** and their **bases 'clasp' the stem** (their sides fold inwards as if to protect the stem).

Turnip *B. rapa* is like Oil-seed Rape but the **lower leaves are hairy and not greyish. Introduced**. By **rivers**, along **roadsides** and on **disturbed and waste ground**.

Black Mustard *B. nigra* has **lower leaves not greyish** and **upper ones** greyish but **stalked**. Native: **riversides, coasts** & **waste ground** but scarce in N Britain & in Ireland.

Wild Cabbage *B. oleracea* is like Oil-seed Rape but **leaves thicker (>0.5 mm thick)**; **lower ones smell of cabbage**; **upper ones only slightly clasp the stem**. Uncommon possible native on **basic soils on sea cliffs**; also a **garden escape on disturbed & waste ground**. Cultivated races include cabbage, sprouts, cauliflower & kale.

Charlock *Sinapis arvensis* has **all leaves hairy, and stems mostly hairy too**. Stem leaves are unstalked. **Introduced**. Along **roads, railway lines** and on **waste ground**.

White Mustard *Sinapis alba* is **hairy**, like Charlock, but **stem leaves stalked**. **Introduced**. On **roadsides** and on **disturbed or waste ground**. Mainly in lowland England.

Common Field-speedwell *Veronica persica*

Low and creeping, with hairy stems and **bright green, short-stalked, oval-triangular leaves with toothed edges** and hairy undersides. Pale blue flowers can occur at any time of year. The **lowest leaves are in opposite pairs**, giving some resemblance to Wood and Germander Speedwells (page 236), but the **upper leaves (with the flowers) are not opposite** (and occupy much of the total stem length on many plants - see photo below). Stem hairy all round (unlike two lines of hairs in Germander S.). Leaf stalk <7 mm long (up to 15 mm in Wood S.) Habitat completely unlike that of Germander and Wood S.: **dryish fertile soils** on **open cultivated and waste ground**. **Introduced.**

Plants in May

Primrose *Primula vulgaris*

Primrose has **rosettes** of long **oval leaves** up to 15 cm long, widest above the middle and **tapering gradually to unstalked bases**. All leaves are at the plant base and have rather **wrinkled uppersides** and felty undersides. In spring the large, **pale yellow flowers** 2–3 cm across are unmistakeable. Leaves of Cowslip (page 271) are more abruptly tapered into their stalks. Leaves of Foxglove (page 272) can look similar but are stalked, more pointed, widest at the middle and grow on the stems as well as at the plant bases; they can also be larger. Primrose grows on **moist, acid to neutral soils** in **woods** (esp. W8–11), **scrub** (W21–22), **hedgerows**, **cliffs** (e.g. U17) and **short, grazed, unimproved grasslands** (e.g. MG2/5, U4 & CG10–12, esp. on N-facing slopes).

Plant in April

Two **smaller** relatives have **greyish leaves** and **yellow-centred pink or purple flowers in clusters at stem tips**. **Bird's-eye Primrose** *P. farinosa* has **pink flowers** on stems up to 15 cm tall and grows in **base-rich flushes** in **N England**, mainly in upland areas. **Scottish Primrose** *P. scotica* is **smaller** (stem up to 10 cm) has **purple flowers** and grows in **short coastal grassland** in **N Scotland** (Sutherland, Caithness and Orkney).

Scottish Primrose

Cowslip *Primula veris*

Cowslip is related to Primrose (page 270), but differs in that its **leaves are shorter and their bases more strongly tapered into stalks**, and the pale yellow **flowers are smaller (<15 mm across) and in clusters at the tips of the stems** (which are <30 cm tall). It grows mainly on **dry to moist, neutral to basic soils** in **short to medium-height** **grazed or mown, unimproved** **grasslands** (esp. MG3-5 & CG1-2 & 8). Also in **taller, less grazed** **grasslands** (MG1 & CG3-6). Commonly included in **newly planted or sown grasslands along roadsides**.

Plant in April

Oxlip *P. elatior* looks like Cowslip but has **paler, larger, more Primrose-like flowers up to 18 mm across. Flowers 'nod' to one side**. Local in **ancient woods** (W8/10) on neutral to basic soils in SE England (mainly East Anglia).

False Oxlip *P. veris x vulgaris = P. x polyantha* looks like Oxlip but the flowers stick out in all directions. Widespread in **woodland, scrub, grassland and by hedges**. Rare in upland Scotland and in NW Ireland.

Plant in a woodland clearing in June

Foxglove

Digitalis purpurea

This species grows **up to 1.5 m tall** and is **unmistakeable in flower**. It is also identifiable by the **oval leaves** up to 30 cm long, on **short, winged stalks** and with **softly hairy, rather wrinkle-textured uppersides** and **grey downy undersides**. Leaves grow at the **base** of the plant and up the **stems**. On **dry to moist, acidic soils** in **open woods** (esp. W10-11), along **hedgerows**, on **cliffs** and on **disturbed ground**. It can be abundant for a few years in **recently felled woodland** or on ground that has been **burned**. All parts of the plant are **poisonous**.

Comfrey leaves (page 274) are rather similar but untoothed. Primrose leaves (page 270) are blunter, widest nearer the tip, more gradually tapered to the base and are mostly smaller (<15 cm long).

Two more **tall woodland** herbs: **leaves are more sharply toothed** and **flowers are pale blue**; mainly on **neutral to base-rich soils**.

Giant Bellflower *Campanula latifolia* has **tapering leaf bases**; mainly in **dampish woodland** (W7/8/9) in central + N Engl., E Wales and S + E Scotl.

Nettle-leaved B. *C. trachelium* has **abruptly-tapering or heart-shaped leaf bases**; mainly in **drier woodland** (W8/12) in central and S England, E Wales and SE Ireland.

Great Mullein *Verbascum thapsus*

Unmistakeable with its **tall thick stems** (up to 2 m) and **large oval leaves** all with a **dense cover of whitish hairs** giving the plant a pale greyish tinge. In summer the **spikes of yellow flowers** are also distinctive. Leaves like Foxglove (p. 272) but with dense woolly cover of hairs. On **dry neutral to basic soils** in **well-lit** places in the lowlands, especially on **disturbed ground** (waste ground, roadsides, railway banks etc), along **hedgerows** and among **scrub** (W21-23).

Plants in July. Flower spikes are typically straight but those on the left have an unusual 'question-mark' form symbolic of its uncertain origin.

Weld *Reseda luteola*

A hairless plant **up to 1.5 m tall**, with **narrow oblong leaves**, commonly with **wavy edges**, and **narrow spikes of small creamy-coloured flowers**. **Introduced**. Grows on **dry neutral to basic soils in the lowlands**, mainly in **disturbed habitats** such as roadsides and waste ground.

Note: The related **Wild Mignonette** *R. lutea* is similar but smaller and with **leaves cut deeply into narrow lobes** - see page 172.

Comfreys *Symphytum* species

Tall plants with **oval, untoothed, hairy leaves**: hairs especially noticeable on margins and along underside of central vein. Leaves are like those of foxglove but without teeth. Curved spikes of flowers at stem/branch tips.

Common Comfrey *Symphytum officinale* has **branched stems up to 1.5 m tall**. **Upper surface of leaf has rather sparse, straightish hairs**. **Base of stem leaf continues as long 'wing' down stem** at least as far as the next leaf down. Lower leaves are up to 25 cm long, their bases tapering abruptly into their stalks (more gradually tapering in the other species). **Flowers are blue or purple**. Native on **moist, more or less neutral soils, especially by watercourses and in mires/fens**.

Russian Comfrey *Symphytum x uplandicum* is like Common Comfrey but **stem leaves continue as 'wings' for only a short distance** down stems (not as far as next leaf down), and lower leaves taper gradually to their stalks. On **drier ground including hedgerows** and **waste ground**. **Not native.**

Tuberous Comfrey *Symphytum tuberosum* has **shorter stems** (to 60 cm), branched only at their tops. **Upper surface of leaf has abundant tiny hooked hairs**. Leaves <15 cm long. Lower leaves taper gradually to their stalks. Stem leaves do not continue as 'wings' down stems. **Flowers creamy yellowish**. Native in **northern British lowlands**, forming **patches** on **moist, more or less neutral soils** in **woods** (esp. W8-11) and along **hedgerows** and **stream/river banks**.

Tuberous Comfrey in flower (May)

Russian Comfrey in flower (May)

Tuberous Comfrey leaves (April)

Common Comfrey: base of stem leaf (June)

Viper's-bugloss *Echium vulgare*

Distinctive with its **very rough, bristly hairs on the stems and leaves**, and **tall flower spikes**. The **leaves are narrow oval**, the basal ones in **low rosettes**, their **upper surfaces with pale raised dots and rough hairs**. The **stems have dark spots and rough hairs**. Flowers 10–18 mm across, in mid to late summer; pink at first; then **blue**. On **dry, neutral to basic soils**, in **grassland** (U1, CG1-2/7, MG1, MC4-5/11), **sand dunes** (SD6-9/19) and on **disturbed ground** e.g. waste ground, field margins & roadside banks). Widespread in **lowland** Britain. Rare N of central Scotland and in Ireland (except E coast areas).

Bristly Oxtongue *Helminthotheca echioides* is also tall with **rough leaves** with **distinct white 'lumps'**, but **flowers are yellow** and **not in spikes**. **Introduced** on **disturbed ground in lowlands**. **Hawkweed Oxtongue** *Picris echioides* is similar but **lacks the white 'lumps' on leaves**. Native in **lowland calcareous grasslands** and **disturbed habitats** in England and Wales.

Bristly Oxtongue

Plant in June

Basal leaves in March

Common Knapweed *Centaurea nigra*

The leaves are **narrow oval, pointed** and mostly 5–20 cm long; the lower ones are stalked; those on the stems unstalked. They are **hairy** on both sides, slightly **roughish**, and their edges untoothed or with a few small teeth or even a few small lobes toward the base. The branched stems (mostly 30–80 cm tall) are quite **stiff, grooved and hairy**. **Purple-pink flowers** in summer. Grows on **dry to moist, neutral soils**, in **unimproved grasslands** (MG1–5) and **roadside verges**. **Palatable** to grazing animals.

Leaves lack net-veined pattern on underside of Devil's-bit Scabious (p. 251) leaves and are less soft and less hairy than those of Red/White Campions (p. 250) & Forget-me-nots (p. 288). Much commoner than Greater Knapweed (p. 164) whose leaves are brighter green, less hairy and more lobed, and whose flower heads have spreading outer 'rays' and different scale patterning (see drawing). Saw-wort (p. 164) has Knapweed-like flowers and branched stems but its leaves are deeply lobed with finely-toothed edges.

Plant in flower in July

Old flower heads in winter

Basal leaves in April

Old dead flower heads in winter

Melancholy Thistle *Cirsium heterophyllum*
Meadow Thistle *Cirsium dissectum*

These are **non-prickly** thistles whose **leaves are oval** with **small weak spines along their edges** and **whitish felty or cottony undersides**. The **deep pink flower heads on tall stems** are conspicuous in summer.

Melancholy Thistle is the larger of the two, with stems up to 1.2 m tall, **lower leaves 20-40 cm long with hairless uppersides** (but young leaves can have hairy uppersides), **toothed edges and side veins closely-spaced in a herringbone pattern**, and **flower heads at least 3.5 cm across. Some plants have leaf margins cut deeply into lobes.** On **damp neutral to basic soils** from Mid Wales and the Peak District northwards (but very rare in Ireland): mainly in **unimproved grasslands** (esp. MG3 & M25), **rush mires** (M23), **Meadowsweet fens** (M27) and **woods** (esp. W7 & W9).

Meadow Thistle is smaller: up to 80 cm tall, with **lower leaves 12-25 cm long with hairy uppersides, toothed or untoothed edges**, and **flower heads up to 3 cm across**. On **damp acid to neutral soils** in Ireland and in Great Britain from Mid Wales, English Midlands and E Yorkshire southwards: mainly in **unimproved grasslands/fens** (esp. M24).

Melancholy Thistle (July)

Melancholy Thistle (June)

Meadow Thistle (October)

Goldenrod *Solidago virgaurea*

Variable (hairy or hairless, branched or unbranched and short to 1 m tall), with **oval leaves (untoothed or with small, well-spaced teeth)** and **spikes of yellow flower-heads**. Basal leaves stalked; stem leaves unstalked. On **dry, acid to neutral soils** in **woods** and **rocky habitats**; commonest in upland areas.

Stem leaf in August

Basal leaves in May

Plant in August

Common Fleabane *Pulicaria dysenterica*

Branched, hairy stems up to 60 cm tall with **oblong-oval untoothed leaves with dense covering of short soft hairs**, and **yellow, Ragwort-like flower heads** 1.5-3 cm across. **Stem leaves have heart-shaped bases 'clasping' the stems**. On **wet acid to basic soils** in **mires/fens** (including M6, M23-25 & M27) and **by water**; common in England & Wales, less so in Ireland and very rare in Scotland.

Young plant in June >

Plants in August

Plants in August

Rosebay Willowherb *Chamaenerion angustifolium*

Distinctive **tall** stems (to 2 m) with **narrow oval-oblong, untoothed, pointed leaves** with hairy uppersides and **grey-green, hairless undersides**. Conspicuous **spikes of pink flowers** in mid to late summer, becoming **fluffy seed pods in autumn**. Dead stems persist through winter. Stems can form **dense patches** (OV27). Grows on **dry to moist, more or less neutral soils**, mainly on **disturbed but not heavily grazed or trampled ground** such as in places where there has been **tree-felling or burning**; also **along railways and roadsides**. It is a **good indicator of artificial disturbance** but it also grows less commonly in its natural habitat of **cliffs**.

Leaves & flowers in August, and (lower right) plants with fluffy seed pods in September

Wood Spurge *Euphorbia amygdaloides*

A distinctive and unusual-looking plant whose upper and lower stems are so unlike each other that they look as though each could belong to a different species. The lower parts of the stems grow in the first year and are thick and **reddish**, with **dark, evergreen, untoothed, narrow oval leaves** up to 8 cm long, widest beyond the middle and **sticking out in all directions**. These leaves are rather clustered toward the top of the first year's growth. The first year shoots grow in loose tufts and are up to about 40 cm tall; they look like seedlings of Rhododendron (page 42) but with leaves smaller, blunter, more varied in size (some very small) and more gradually tapering at the bases into their very short stalks. The upper part of the stem grows in the second year and can reach about 80 cm above ground level. It is greener and slightly hairy, with well-spaced small, inconspicuous oblong leaves and **large, branched flower heads** with **many roundish, yellowish flowers 1–2 cm wide**. Wood Spurge grows on **dry, acid to neutral soils** in **lowland woods** (esp. W10) and **hedgerows** in **S England** and (less commonly) **Wales**.

Plant in April

Plants in June

Mouse-ear Hawkweed *Pilosella officinarum*

A **short** plant with **creeping above-ground shoots (runners)** and **oval, untoothed leaves** with **very pale undersides and long white hairs scattered on upper and lower surfaces**. Lowest leaves are in small **rosettes**. Leaves on runners are smaller. Rosettes and runners together can form **low mats**. **Yellow, Dandelion-like flower head** in summer, on **upright hairy leafless stem** 10–20 cm long. Flowers paler yellow than in Dandelion. Grows on **dry, acid to basic soils**, mainly at low altitude, in **short, grazed calcareous grasslands** (CG1–2 & 7–10), **acid grasslands** (U1 & 4) and **sand dunes** (SD7–8 & 11–12). Also in taller, lightly grazed lowland calcareous grassland (CG3–6) and on **rocky ground** where rockiness prevents larger plants from taking over. Appears **unpalatable** to grazing animals such as sheep, cattle and deer.

Flowering plant in June

Fox-and-cubs *P. aurantiaca* has similar leaves but differs in its unmistakeable **clusters of orange flower heads** at the shoot tips. **Not native**; a commonly-naturalized garden escape in **grasslands** and **disturbed habitats**; especially common in or near urban areas.

Hawkweeds *Hieracium* agg.

A aggregate of about 420 microspecies among which separation is a difficult specialist task that even most botanists don't attempt. Most Hawkweeds have **oval leaves** growing in a **rosette at the base** of the plant **and up the stems**, but some species lack a basal rosette and some others have very few stem leaves. The leaves and their stalks are **hairy** to some degree, and typically have **small to medium-sized teeth spaced quite far apart** along their edges. The **stems are 30–100 cm tall, branched** in their upper parts and bear **yellow, dandelion-like flower heads** 2-3 cm across. Mouse-ear Hawkweed (page 281) differs in its untoothed leaves, unbranched flower stems and above-ground creeping shoots (runners). Cat's-ear and Hawkbits (page 166) have narrower, lobed basal leaves and no stem leaves. Smooth Hawk's-beard (page 166) has hairless, narrower, lobed basal leaves. Marsh Hawk's-beard (page 283) also differs in being hairless, and Sow-thistles (p. 167) have hairless, sharply lobed, prickly-looking leaves. Hawkweeds grow in many habitats but are commonest on **dry, rocky ground in well-lit places**.

Flowering stem and lower leaves in September

Marsh Hawk's-beard *Crepis paludosa*

A rather tall (50–100 cm), **hairless** plant whose **oval leaves have large backward-pointing teeth** and whose branched stems bear **yellow, dandelion-like flower heads** up to 3 cm across. The lowest leaves are on **short stalks**; those up the stems are **unstalked** and their bases grow 'backwards' beyond the stem. Grows on **damp to wet, neutral soils** in **woods** (esp. W7 & 9), **mires & fens** (esp. M23, 26 & 27), **grasslands** (U4–6) and **springs** (M37). **Palatable** to grazing sheep, deer and cattle; grows **mainly where not heavily grazed**.

Flowers in June

Leaf at base of plant (these leaves have stalks)

Stem leaves (these are unstalked)

Nipplewort *Lapsana communis*

A tall (50–90 cm) plant with **branched stems** that are hairy in their lower parts but hairless higher up. Upper branches are very slender. The **leaves** are **hairy**: those on the stems are **oval with large, widely-spaced teeth** (uppermost leaves can be smaller, very narrow and less toothed); lower leaves are similar but with their **bases cut deeply into a few lobes** on either side (lobe margins cut almost to central vein of leaf). The **flowers**, in summer and early autumn, are **yellow and dandelion-like but smaller** and up to 2 cm across: there can be many flowers on one well-branched plant. Grows mainly on **dryish, neutral soil** in **woods** (esp. W8–10 and **woodland margins**), **scrub** (esp. W21), **hedgerows, roadsides and disturbed or waste ground**. Mainly in the **lowlands**.

Plant in flower in June

Old seed heads in January

Leaves and lower stems in June

Daisy *Bellis perennis*

Low rosettes of **small, short-stalked, oval**, shortly-hairy **leaves** with shallowly-toothed edges. **White and yellow flower heads** (1.5-2.5 cm across) on short leafless stems. Mainly on **moist, neutral to basic soils**, in **unimproved or improved grasslands** (esp. MG5-7) kept **short** by **grazing, mowing or trampling**. Also in **disturbed habitats** such as **roadsides** and **waste ground**.

Plants in April

Oxeye Daisy *Leucanthemum vulgare*

The hairy, rather stiff stems up to 70 cm tall are topped by **white flower heads like those of daisy but larger**: 2.5–6.0 cm across. Leaves at base of plant are **oval to rounded** (like daisy leaves) and on stalks. The **leaves growing up the stems are unstalked and are oval with large blunt teeth** (that can be almost like small lobes) along their edges. Mainly on **dry, neutral to basic soils**, in **unimproved, grazed or mown grasslands** (esp. MG5) and on **disturbed** ground.

Flowering plants in neutral grassland in June

Toothed leaves on stem

Roundish leaves at base of plant

Harebell *Campanula rotundifolia*

Often called **Bluebell** in Scotland. A delicate, **hairless** plant with **thin stems** and unmistakeable **pale blue, bell-like flowers** in mid to late summer. Leaves are all small and inconspicuous. **Lowest leaves are on long stalks and are roundish** with a few teeth. Up the stem they become **very narrow, untoothed and unstalked**. Harebell grows mainly on **dry, acid to basic soils**, in **short** (e.g. **grazed**) **unimproved grassland** (U1/4, MG5, CG1-2 & 7-10).

◄ Left: flower in July

▼ Below: leaves in June

Lower photos: lower leaves in July. Lower leaves can be Violet-like but teeth fewer, more spaced and more irregular-sized, and side veins more branched and less strongly curved.

Forget-me-nots *Myosotis* species

The **narrow oval leaves** are **blunt or pointed, soft-textured** and **hairy on both sides**. The **basal leaves are short-stalked. Leaves on stems are unstalked. Side veins of leaves are faint**. Leaves are smaller, softer & blunter than in Common Knapweed (page 276) and Devil's-bit Scabious (page 251), and smaller and narrower than in Red & White Campions (page 250, whose leaves are in opposite pairs). Slender upright stems are 5-40 cm tall (<10 cm in Changing & Early F.), with spikes of **small, pale blue flowers** at their tips. The commonest species are described below:

IDENTIFIC-ATION	**Hairs on lower part of stem**	**Hairs on upper part of stem**	**Max. leaf length**	**Leaf width**	**Flower diam.**
Field F. *M. arvensis*	sticking outwards	sticking outwards	6.5 cm	7-15 mm	3-5 mm
Wood F. *M. sylvatica*			8 cm	20 mm	6-10 mm
Water F. *M. scorpioides*		pressed against stem	10 cm	10-20 mm	4-10 mm
Creeping F. *M. secunda*			5 cm	5-15 mm	4-6 mm
Changing F. *M. discolor*			4 cm	4-7 mm	2 mm (yellowish at first)
Early F. *M. ramosissima*			1.5 cm	3-6 mm	2 mm
Tufted F. *M. laxa*	pressed against stem		8 cm	10 mm	2-4 mm

HABITATS

Field F. *M. arvensis*: **dry neutral soils** in **disturbed field margins** & **hedgerows**. **Not native.**
Wood F. *M. sylvatica:* **moist neutral soils** in **lowland woods**. **Native + planted/naturalised.**
Water F. *M. scorpioides* & **Tufted F.** *M. laxa*: **wet neutral-basic soils**, in **herb/grass/rush mires & by water**. Tufted F. esp. where **disturbed**. Both species are native.
Creeping F. *M. secunda*: **wet acid soils**, in **herb/grass/rush mires**. Native.
Changing F. *M. discolor*: **open** or **disturbed dry grassland** at low altitudes. Native.
Early F. *M. ramosissima*: **open** or **disturbed dry grassland** in lowlands. Native.

Field Forget-me-not leaves (October) and flowers (May)

Navelwort *Umbilicus rupestris*

Also called **Wall Pennywort**. Unmistakeable **round, wavy-edged leaves** are **smooth, shiny, thick, fleshy-textured and evergreen**. Leaf veins not obvious. Leaf stalks attached to centre of underside of leaf. **Creamy flowers in upright spikes** in summer, remaining (brown) through winter. On **acid to neutral rocks** and **walls** (less commonly on tree trunks/bases) in **SW half of Britain** N to Mull and Ardnamurchan, and in **Ireland**. Grows in **woods** (for example on/among rocks in W8-11 & W17), in **hedgebanks** and on rocks & walls in more **open situations**.

Plant in April

Marsh Pennywort *Hydrocotyle vulgaris*

This species has round leaves like those of Navelwort but **not thick, fleshy or evergreen, and with narrow but distinct veins**. Leaf stalks and leaf undersides can have scattered hairs. Small inconspicuous clusters of tiny pale green-pink flowers in summer. Grows on **wet, neutral soils**, in **mires/fens/swamps** (esp. M5/13/22-29 & S12/19/21/24/25/27) and **dune slacks** (SD13-17).

Leaves in November

Small insectivorous plants of wet habitats

Small plants that grow in nutrient-poor habitats, especially in the N & W, and supplement their food source with nutrients from insects caught on their leaves.

Sundews *Drosera* species

Low rosettes of small, stalked, reddish leaves fringed with sticky hairs. The leaves sense an insect landing and fold in to catch it. The flowers are small and white, in spikes on upright stems. Sundews grow in **bogs** (M17–18), **wet heaths** (M15–16), **flushes** (M10–11 & M15a) and on **loch shores**. **Round-leaved Sundew** *D. rotundifolia* is by far the commonest and is easily identified by its **round leaf blades**. **Oblong-leaved Sundew** *D. intermedia* has **narrow oval leaf blades c.1 cm long**, and a **flower stem starting below the leaf rosette** and then curving out and up; it grows mainly in the west. **Great Sundew** *D. anglica* has **narrow oval leaf blades c.3 cm long** and a **flower stem growing straight up from the middle of the leaf rosette**; it grows mainly in the N and W, especially in **wet bogs**.

Butterworts *Pinguicula* species

Low rosettes of sticky, pale leaves with upturned edges. Insects get stuck on the leaves. **Common Butterwort** *P. vulgaris* has **yellow-green leaves 2–8 cm long**, and **purple flowers c.1.5–2 cm across** on upright stalks; on **wet, neutral to basic muddy or peaty soils, especially in flushes** (M10–11, M15a) where **vegetation is kept short or sparse** by **wetness, rockiness** or **grazing/trampling**. **Large-flowered Butterwort** *P. grandiflora*: similar but with **bigger flowers 2.5–3.5 cm across**; in similar habitats; in **SW Ireland**. **Pale Butterwort** *P. lusitanica*: mainly western; less common than Common Butterwort, but the commoner of the two in S/SW Engl.; **smaller, duller pale brown-green leaves <2 cm long; whitish-pink flowers**; in **flushes** (esp. M10).

Round-leaved Sundew leaves in June

Common Butterwort in flower in May

Ribwort Plantain *Plantago lanceolata*

Low rosettes of **narrow, dull or greyish-green, hairy leaves** with several **parallel veins** running the length of the leaf. **Leaves taper very gradually to their narrow bases and very short stalks**. **Short**, dense **blackish** flower heads (with creamy stamens) on leafless stems. On **dryish neutral to basic soils** in **unimproved grasslands** (MG1/3-5/8-12, CG1-11 & U1/4 & 5), in **improved grasslands** (MG6-7), along **hedgerows** and **roadsides**, and on **waste ground**.

Greater Plantain *Plantago major*

Has **broader leaves** that are **hairless or with short hairs** and taper more **abruptly** into **distinct leaf stalks** that can be as long as the leaves themselves. Flower head **long**, narrow and **green** (brown when dead). On **dry to moist neutral soils**, especially where **trampling** keeps **vegetation short & open**: **hedgerows, roadsides, paths, field margins** and **waste ground**.

Ribwort Plantain in April

Greater Plantain in April

Hoary Plantain *P. media* has **broad leaves** and **long flower spikes** as in Greater Plantain, but the **leaves are distinctly hairy** (can look rather **greyish-green**) and with **hardly any leaf stalk**. Mainly on **dry basic soils** in **lowland unimproved calcareous grassland** (CG1-8) in England and Wales. Rare in S Scotland. Scarce introduction in N Scotl. & in Ireland.

Thrift, Sea Plantain & Sea Arrow-grass

These are among the commonest herbs in **saltmarshes**. Thrift and Sea Plantain also grow on **sea cliffs** and in **rocky places in hills**. Easily identified in flower, these species can be confused at other times because they all have **low, dense tufts of hairless, long, narrow, parallel-sided leaves**.

Thrift *Armeria maritima* has **pink flower heads** and its **leaves are 1-3 mm wide** with a **central groove along on the upper surface**. **Coastal rocks, grasslands, heaths and saltmarshes**. Also **mountain rocks, grassland and flushes**.

Sea Plantain *Plantago maritima* has **dense, green flower spikes up to 5 cm long** and its **thick leaves are mostly about 3-8 mm wide**, with **no more than a very shallow, indistinct groove** down the middle. **Coastal rocks, grasslands, heaths & saltmarshes**, and **upland rocks & flushes**. Differs from Buck's-horn Plantain (page 170) in its hairless leaves with no lobes and few or no teeth.

Sea Arrow-grass *Triglochin maritima* has **longer (5-15 cm) spikes of small green flowers** and its **leaf is mostly 2-3 mm wide, half-cylindrical in section** and with a **wide, sheathing base at the top of which is a small, narrow projecting bit (ligule) up to 6 mm long** (no ligule or sheathing leaf base in Sea Plantain). Common in **saltmarshes**. (**Marsh Arrow-grass** *T. palustre* is **more slender, with leaves mostly <2 mm wide and ligules <1 mm long**. In a wide range of vegetation types on **wet, non-saline, neutral to basic soils**.)

Thrift in flower (June)

Thrift leaves (June)

Sea Plantain (June)

Sea Arrow-grass flower spikes & leaf bases (Sept)

Small Stonecrops *Sedum/Petrosedum*

Small plants with **small oval leaves that are very thick in relation to their size**. Stonecrops grow mainly on **thin soils among rocks and on walls**.

IDENTIFICATION	Leaf length	Leaf tip	Old dead leaves on lower stems	Leaves	Flower colour
English Stonecrop *S. anglicum*	3–5 mm	blunt	fall off	Green to reddish	White
Biting Stonecrop *S. acre*	3–5 mm	blunt	remain		Yellow
White Stonecrop *S. album*	6–12 mm	blunt	fall off	Brightish green	White (in dense cluster)
Hairy Stonecrop *S. villosum*	6–12 mm	blunt	remain	Reddish + short hairs	Pink
Reflexed Stonecrop *Petrosedum rupestre*	10–20 mm	with tiny point	fall off	Greyish-green	Yellow (in dense clusters)
Rock Stonecrop *P. forsterianum*	10–20 mm		remain		

HABITATS	
English Stonecrop *S. anglicum*	Acid rocky habitats; mainly in W
Biting Stonecrop *S. acre*	Acid to basic rocky habitats + dunes; widespread
White Stonecrop *S. album*	Neutral to basic rocky habitats + walls, roofs etc; introduced and widespread
Hairy Stonecrop *S. villosum*	Neutral to basic flushes (especially stony ones) in uplands of N England, S Scotland & SE Highlands.
Reflexed Stonecrop *P. rupestre*	Rocks, banks & walls; introduced & widespread
Rock Stonecrop *P. forsterianum*	Rocks & screes; native in parts of Wales & SW England; introduced elsewhere (but not common)

English Stonecrop (July)

Biting Stonecrop (June)

Roseroot *Rhodiola rosea*

Conspicuous and unmistakeable with its **thick, hairless, unstalked, grey-green leaves** up to 4 cm long and toothed at least toward the tips, **sticking out in all directions around the thick stems** that are up to 30 cm long. Brown scales at base of stem. Clusters of small yellow flowers at shoot tips in summer. Grows in the **N & W** of Britain & Ireland, on **moist, ungrazed cliff ledges** in the **uplands** (U17) and on **sea cliffs** (MC3). A good **indicator of basic soils**.

Roseroot plants in July

Orpine *Hylotelephium telephium* looks similar but has **red flowers**, and **no brown scales at the stem base**. On **dry, neutral to basic soils** in **little-grazed dry banks, hedgerows** and **woodland margins** in **lowlands**. Native except in Ireland (scarce) and Isle of Man.

Sea Aster *Tripolium pannonicum*

Another plant with **thick, hairless, grey-green leaves**, these being **narrow oval or oblong, blunt-tipped, untoothed** and **up to 12 cm long**. Lower leaves are stalked; those on the stems unstalked. In summer-autumn: **branched groups of pale blue-violet flower heads up to 3 cm across, with yellow centres** (just the yellow centres on some plants). Common in **saltmarshes**.

Sea Aster in September

Michaelmas-daisies *Symphyotrichum* species have flowers like Sea Aster but **toothed or hairy, green (not grey-green) leaves**. **Introduced**: on **disturbed/waste ground** and **riversides**.

Sea-purslane *Atriplex portulacoides* has **fleshy grey-green narrow oval leaves** like Sea Aster but shorter (up to 5 cm long). It is **shrubby** with branched stems (with **woody lower parts**) up to 80 cm tall, or low and creeping. **Lower leaves are in opposite pairs**. It has spikes of **tiny dull greenish flowers**, as in Oraches (page 265). It grows on **saltmarshes** (esp. where **ungrazed**) in England, Wales, E Ireland and Solway Firth coast of Scotland.

Lesser Spearwort *Ranunculus flammula*

This **hairless** plant is mostly <40 cm tall, with **yellow flowers 1-2 cm across**, like those of **Buttercups** (pages 200-202) but with slightly **narrower petals**. The **leaves** are up to 5 cm long, **narrow oval, pointed, slightly toothed and hairless**, with side veins more or less **parallel** to the central vein. The lowest leaves are on **long stalks**; leaves on the stems are **short-stalked or unstalked**. Stems can be **red-tinged**. rows on **wet, neutral to moderately acid soils** in **mires and flushes** (M5, M22-23, M29 & M35), in **dune slacks** (SD14-17), and on **muddy or stony ground along lakeshores**. **Unpalatable** to grazing animals.

Leaves in June

Flowering shoots in June

Flowering plants in rush mire in June

Greater Spearwort *R. lingua* is similar but **larger**: **up to 120 cm tall**; **leaves up to 25 cm long**; **flowers 2-5 cm across**. Rather uncommon (as both a **native** and an **introduction**) on **very wet, neutral to basic soils** in the **lowlands**, in **fens**, and in **swamps in shallow water**.

Wild Garlic *Allium ursinum*

Distinctive **large, long oval, smooth** (hairless), **bright green leaves with numerous parallel veins**, up to 7 cm wide, tapering gradually to **pointed, drooping tips** and tapering at bases to **long stalks**. Clusters of **white flowers** on long, leafless stalks. Whole plant **smells of garlic**. Leaves are **edible**: they taste of garlic. Leaves can carpet the ground in spring, but die away by winter. Grows on **deep, moist, neutral soils** in **woods** (W7–9). Commonest in **ancient woods**; most restricted to this habitat in drier southern lowland areas.

Wild Garlic in flower May

Two related **introduced** species with **narrow, parallel-sided leaves <1.7 cm wide**, in **lowland woods, parks, riversides and disturbed and urban habitats**; both **can be invasive**.

Few-flowered Garlic *A. paradoxum.* **Leaves are flat in section**. Flower head has **green bulbils** and just a **few (or no) flowers**; very pale **bracts** at base of flower head. Mainly in E Britain and E Ireland.

Three-cornered Garlic *A. triquetrum.* **Leaves are triangular in section** (projecting 'keel' on underside). **Flower head lacks bulbils and bracts** and has **several to many drooping flowers**. Widespread in S half of Britain and in Ireland. Rare in Scotland.

Few-flowered Garlic in April

Bluebell *Hyacinthoides non-scripta*

Also called **Wild Hyacinth**. Has tufts of **hairless leaves that are long (20–40 cm), narrow (7–15 mm), thick, parallel-sided and blunt-tipped**, and can form carpets in woods. Daffodil leaves (page 298) are greyer in colour. Snowdrop leaves (page 298) are shorter (<20 cm), narrower (4 mm) and greyer. Greater Woodrush leaves (page 147), which can also form carpets in woodland, taper to pointed tips and have long white hairs along their edges. In spring Bluebell has **one-sided, droop-tipped spikes** of **hanging blue flowers that are narrow and parallel-sided, with reflexed tips**. Old flowers point upwards when in fruit. Dead, pale brown fruiting stems remain into autumn. Native on **moist, mildly acid to mildly basic soils** in **woods** (esp. **ancient** W8–11), under **bracken** (W25 & U20) and on **cliffs**. Mainly in **ungrazed to moderately grazed** places. In drier southern lowlands it is most strongly restricted to ancient woodland.

Hybrid Bluebell *H. hispanica* x *non-scripta*: similar but has **wider leaves (to 3.5 cm wide)** and **straighter, more upright spikes** of **flowers that stick out more widely from the stem and widen gradually to less reflexed tips**. **Naturalised** in **woods**, mainly in **lowlands**; esp. near gardens, villages, towns etc; might threaten purity of native Bluebell populations.

Bluebells in West Highland woodland in May

Bluebell: flower in May (L); old flower head in autumn (R)

Hybrid Bluebell (May)

Snowdrop *Galanthus nivalis*

This tufted plant is **unmistakeable when in flower in late winter and early spring**. It can also be told by the leaves: **short** (mostly 10–20 cm), **narrow** (4 mm), **thick, grey-green, blunt-tipped and hairless**. **Non-native: planted** or **naturalised**, mainly on **moist, neutral soils in lowland woods**. Daffodil leaves (p. 298) are longer (to 35 cm) and wider (to 12 mm). Bluebell leaves (p. 297) are brighter green (not grey-green), longer (to 40 cm) and wider (to 15 mm).

Plants in February

Daffodil *Narcissus pseudonarcissus*

Tufts of **hairless grey-green leaves** 20–35 cm long and up to 12 mm wide, and in early spring unmistakeable **large yellow flowers** on long stalks, each flower with pale outer petals and a darker trumpet-shaped central tube. Leaves longer and wider than those of Snowdrop (page 298); greyer than in Bluebell (page 297). Mainly on **moist, more or less neutral soils**: native in **woods** in most of England and the Welsh Borders; **planted and naturalised elsewhere**.

Bog Asphodel *Narthecium ossifragum*

Flattened tufts of **short, smooth, curved leaves** 5-20 cm long and 2-5 mm wide (like a miniature Yellow Flag), and, in summer, **spikes of yellow flowers** (like a yellow-flowered orchid). On **wet, acidic, peaty soils**. Mainly in **short, open** or **grazed wet heaths** (M14-16) and **bogs** (M17-18 & 21). Only **moderately palatable** to grazing animals, probably because it is **rather toxic**.

Old dead flower heads in winter

Leaf tufts and a flower spike in July

Scottish Asphodel *Tofieldia pusilla* is like a miniature version of Bog Asphodel. Its leaves are **smaller** (<4 cm long; <2 mm wide), **straighter**, and have **tiny teeth along one edge**. In summer there is a small dense head of **tiny whitish flowers** on a short, slender stem. It grows in **short, open vegetation** in **wet, base-rich flushes** (M10-11) at moderate to high altitudes in the Highlands and (rare) the N Pennines.

NOTE: Non-flowering Annual Meadow-grass (page 97) with leaves in flattened tufts can look surprisingly similar to Scottish Asphodel but has flattened leaves folded up along each side of the central vein, and some leaves have wrinkled bases.

Scottish Asphodel

Tall plants with flattened tufts of long, parallel-sided, hairless leaves

Yellow Flag *Iris pseudacorus*: **green leaves 15-25 mm wide and up to 1m long** (about as tall as stems); **large yellow flowers** in summer. On **moist to wet, neutral soils** in **mires** (esp. M28) & **by streams/ponds**. **Appears rather unpalatable**; eaten mainly in heavily-grazed places. **Sweet Flag** *Acorus calamus*: **leaves similar but wrinkled**; flower a **thick green lateral spike**; **introduced**; lowland **shallow water**; can form swamps (S15).

Montbretia *Crocosmia x crocosmiiflora* has **leaves up to 2 cm wide and 50 cm long** (shorter than stems); **several parallel veins form ridges or 'pleats' on both surfaces** (Yellow Flag leaves are smoother). Flowers **smaller, orange, and in branched spikes**. A **garden escape**: mainly in **lowland woods, wood margins and hedges**.

Branched Bur-reed *Sparganium erectum* has **leaves 10-15 mm wide; central vein sticks out as a raised ridge running down middle of underside. Branched groups of spherical flower heads** (in summer). **In shallow water**; can form **swamps** (S14); **palatable** to grazing animals. Three related aquatic species, all less common and with **unbranched flower shoots. Unbranched B-r** *S. emersum*: slightly **smaller**; upright + floating leaves <12 mm wide; widespread. **Least B-r** *S. natans*: **small**; **flat floating/submerged leaves <6 mm wide**; mainly NW Brit. + Ire. **Floating B-r** *S. angustifolium*: **long, flat floating leaves <4 mm wide**; mainly in uplands.

Bulrush *Typha latifolia* has **grey-green leaves 12-18 mm wide**, (greyer than Yellow Flag and lacking the ridges/pleats of Montbretia or the raised central vein of Branched Bur-reed) and **cylindrical dark brown flower heads** on thick stems up to 2 m tall. On **wet, nutrient-rich soils**, in **swamps** (S12) and **by ponds & lakes**; mainly in **lowlands**, in **ungrazed** places (too wet for animals to gain access).

Lesser Bulrush *T. angustifolia* is similar but with **thinner leaves 4-5 mm wide**. In **similar habitats** (esp. S21) but mainly in **southern lowlands**.

Yellow Flag

Montbretia

Branched Bur-reed

Bulrush

Common pink/purple-flowered orchids

These have **hairless, narrow oblong or oval leaves** (some with dark spots) and **upright spikes of flowers**. The commonest pink- or purple-flowered species are described on these two pages. Seven more species are on page 303.

Common Spotted

Heath Spotted

Northern Marsh

Fragrant

The habitats of our commonest pink or purple-flowered orchids are described below, and notes on their identification are given on the next page.

HABITATS OF COMMON PINK/PURPLE-FLOWERED ORCHIDS:

Common Spotted-orchid *Dactylorhiza fuchsii*	**Neutral to basic soils in woods, scrub, hedges, unimproved grasslands & herb-rich mires/fens.**
Heath Spotted-orchid *Dactylorhiza maculata*	**Moist, acid soils in heaths, flushes and unimproved grasslands, especially in upland areas.**
Early-purple Orchid *Orchis mascula*	**Neutral to basic soils in woods, scrub, hedges, unimproved grasslands, cliffs and road/railway banks.**
Northern Marsh-orchid *Dactylorhiza purpurella*	**Moist, neutral to basic soils in mires/fens** & **unimproved grasslands** NW of Swansea-Hull line.
Southern Marsh-orchid *Dactylorhiza praetermissa*	**Moist, basic soils in mires/fens** & **unimproved grasslands** in lowlands from Lancashire & Yorkshire southwards.
Early Marsh-orchid *Dactylorhiza incarnata*	**Moist, neutral to basic soils in mires/fens, flushes and unimproved grasslands.**
Fragrant Orchid *Gymnadenia conopsea s.l.**	**Dry to moist, neutral to basic soils in unimproved grasslands, mires and heaths.**
Pyramidal Orchid *Anacamptis pyramidalis*	**Dry to moist, basic soils, mainly in lowland unimproved grasslands and sand dunes**; rare in Scotland.

* = *sensu lato* = in the broad sense: three very similar species (*G. borealis, G. conopsea* and *G. densiflora*) grouped together.

IDENTIFICATION OF COMMON PINK/PURPLE-FLOWERED ORCHIDS:

Pyramidal Orchid

Leaves with dark spots, or unspotted	Stem solid or hollow *	Lower part of stem v. thick?	Colour of flowers	Additional notes	
Spotted (less commonly unspotted)	Solid	No	Pale pink	Leaf >1.5 cm wide; mainly dull green. On each flower, lower petal deeply divided into 3 lobes (middle longest).	**Common Spotted-orchid** *Dactylorhiza fuchsii*
				Leaf <1.5 cm wide; mainly dull green. On each flower, lower petal no more than slightly 3-lobed (middle lobe smallest & shortest).	**Heath Spotted-orchid** *Dactylorhiza maculata*
			Bright pink-purple	Main leaf colour shiny green.	**Early-purple Orchid** *Orchis mascula*
Unspotted (or a few spots at leaf tips)	Hollow	Yes	Deep pink-purple	Lowest petal of flower is less than 9.5 mm wide.	**Northern Marsh-orchid** *Dactylorhiza purpurella*
			Medium pink	Lowest petal of flower is more than 9.5 mm wide.	**Southern Marsh-orchid** *Dactylorhiza praetermissa*
			Mostly dull flesh-pink	Leaf tips 'hooded' (curved up like front end of boat). Flowers can be red, pink or dull yellow.	**Early Marsh-orchid** *Dactylorhiza incarnata*
Unspotted	Solid	No	Bright pink	Leaf tips slightly 'hooded'. Flower has strong, pleasant smell.	**Fragrant Orchid** *Gymnadenia conopsea s.l.*
			Bright pink-purple	Leaf tips slightly 'hooded'. Flower spike short and triangular, with pointed tip.	**Pyramidal Orchid** *Anacamptis pyramidalis*

* Check by gently squeezing the stem; if it's hollow it will feel rather soft and spongy.

Seven more orchids

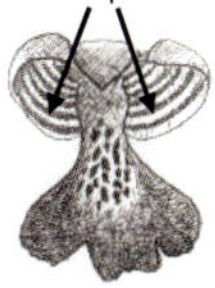

Green-winged Orchid *Anacamptis morio* has **unspotted leaves** and **purple flowers whose upper side parts have conspicuous stripes typically including green**. On **neutral to basic soils** in **unimproved grassland** (esp. MG5) in lowlands of England, Wales & central Ireland; very rare in S Scotland & S & N Ireland.

Bee Orchid *Ophrys apifera* has **unspotted leaves** and relatively few, large **pink flowers with a multi-coloured central part resembling a bumble-bee**. On **dry basic soils** in **unimproved grassland** (CG2-5) and on **disturbed ground** in lowlands of England, Wales and Ireland; extremely rare in S Scotland.

Greater Butterfly-orchid *Platanthera chlorantha* has **two unspotted oval leaves** up to 15 cm long and 5 cm wide **at base of plant**, and **spikes of whitish flowers** >18 mm across Mostly in **unimproved grassland** (esp. MG5) **on dry neutral to basic soils** at low altitudes.

Lesser Butterfly-orchid *Platanthera bifolia* is similar but **smaller** with leaves <9 cm long and <3 cm wide, and flowers mostly <18 mm across. Mostly on **damp acid to basic soils**, in **heaths, unimproved grassland** (esp. MG5) and **among bracken**, at low altitudes.

Common Twayblade *Neottia ovata* has **two broad oval green leaves** (5-20 cm long and 5 cm wide) **on lower part of stem**, and a spike (7-25 cm long) of **small yellow-green flowers** 10-15 cm across. Up to 75 cm tall. On **neutral to basic soils**; mainly at low altitudes in **woodland** (esp. W8), **scrub, dunes** (SD8) and **unimproved grassland** (MG3-5 & CG6/8).

Broad-leaved Helleborine *Epipactis helleborine* also has **greenish (to brownish) flowers** but in **taller, less dense spikes**, and **several broad oval leaves scattered singly well up the stem**. Up to 80 cm tall. On **dry to damp, mainly neutral to calcareous soils,** in **woodland** (esp. lowland W7-10), **scrub** and **disturbed** and **urban** habitats.

Lesser Twayblade *Neottia cordata* is like Common Twayblade but **much smaller**: leaves <2.5 cm long; flower spikes 1.5-6 cm long. On **damp acid soils**, in **heaths** (esp. H12 & H21), **bogs** (M19) and **damp woods** (W4 & 17-18) in upland areas of the N and W.

Section 3f

MOSSES AND OTHER MOSS-LIKE PLANTS

Short plants with tiny leaves or just a simple flat, tongue-like 'thallus'. No flowers. To determine if your plant is a moss, liverwort, filmy-fern or clubmoss, look closely at the leaves: their shape and how they are arranged along the shoots. Use a hand lens if you have one (see next page). Compare what you see with the descriptions below. The next four pages give further illustrated introductory information.

MOSSES (pages 309-331):

- Leaves mostly translucent; oval, triangular or long and very narrow.
- Leaves arise from all sides of shoots - not in two strict rows to left & right.

LEAFY LIVERWORTS (pages 332-335):

- Leaves translucent; bluntly oval, oblong or round, or cut into 2-5 lobes.
- Leaves strictly in two rows along left & right sides of shoot (there can also be a third row of very tiny 'underleaves' along the underside of the shoot).
- See also filmy ferns on page 335 - they look rather like big leafy liverworts.

THALLOID LIVERWORTS (page 333):

- Just a simple, flattened, tongue-like 'thallus'; no separate stems and leaves.

FILMY FERNS (page 335):

- Like large leafy liverworts with deeply divided dark green, translucent leaves.

CLUBMOSSES (page 336) (actually related to ferns):

- Leaves opaque, thick and tough; narrow oval with pointed or blunt tip.
- Stems strong (can seem almost woody); mostly irregularly branched.

Note: Procumbent Pearlwort (page 242) can look moss-like but has flowers (though these are small), and its narrow, pointed leaves are in opposite pairs along the thin stems.

A few things about how these plants look, and how we can look at them

They don't die back in winter, and are not grazed by animals. This means they are fully 'presentable' all year round, wherever they grow.

What looks from a distance like 'a patch of moss' often turns out to be several species growing together to form little 'communities' of mixed colours, textures and growth forms.

The appearance of a moss or liverwort can vary depending on whether it is moist or dry. Leaf details are most easily observed when moist: leaves then tend to spread out more and look brighter-coloured.

A hand lens is useful for looking at small details. A good magnification is x10. A x20 lens shows more detail but can be quite fiddly to use. If you don't have a lens you can magnify things by looking backwards through one side of a pair of binoculars.

A tip for using a hand lens... When looking through a lens at a bit of moss or liverwort held in your hand, it can help to hold the specimen up to or higher than eye level, so it is back-lit by a pale background. As you now look 'through' the translucent leaves some details may show up better. (The only trouble is that your eyes adjust to the lightness, so that afterwards when you look around again and down toward the ground everything can look very dark for a while until you eyes re-adjust!)

Another tip... Collected specimens (take only small bits of course) can be put into a plastic bag to take home for further study. After they have dried out you can keep them indefinitely in paper packets, and you can re-examine them any time and even re-moisten them for this purpose. Store them dry.

The mosses, liverworts, filmy-ferns and clubmosses in this book

In Britain and Ireland we have about 750 species of moss (58 in this book), nearly 300 liverworts (16 in this book), two filmy-ferns (both in this book) and nine clubmosses (five in this book). Mosses and liverworts, known collectively as bryophytes, grow on all sorts of surfaces: ground, rocks, walls, trees, rotting logs... They grow larger and more abundantly in damp and shaded places, especially woods, ravines, bogs, springs, flushes, among rocks and on mountains. The species in this book grow mainly in these habitats rather than drier or artificial ones (such as roadsides, paths and arable fields) whose species tend to be smaller and harder to identify. Mosses and liverworts have Latin and English names. The Latin ones are actually easier, but for completeness the Latin and English names of each moss and liverwort species in this book are given in the list below.

Latin and English names of mosses & liverworts in this book:

Atrichum undulatum: Common Smoothcap
Bazzania trilobata: Greater Whipwort
Brachythecium rutabulum: Rough-stalked Feather-moss
Brachythecium rivulare: River Feather-moss
Calliergonella cuspidata: Pointed Spear-moss
Campylopus atrovirens: Bristly Swan-neck Moss
Conocephalum conicum: Great Scented Liverwort
Dicranum majus: Greater Fork-moss
Dicranum scoparium: Broom Fork-moss
Diplophyllum albicans: White Earwort
Eurhynchium striatum: Common Striated Feather-moss
Frullania species; Scaleworts
Homalothecium sericeum: Silky Wall Feather-moss
Homalothecium lutescens: Yellow Feather-moss
Hylocomiadelphus triquetrus: Big Shaggy-moss
Hylocomium splendens: Glittering Wood-moss
Hypnum andoi: Mamillate Plait-moss
Hypnum cupressiforme: Cypress-leaved Plait-moss
Hypnum jutlandicum: Heath Plait-moss
Hypnum cupressiforme var. *lacunosum*: Great Plait-moss
Isothecium alopecuroides: Larger Mouse-tail Moss
Isothecium myosuroides: Slender Mouse-tail Moss
Kindbergia praelonga: Common Feather-moss
Lepidozia reptans: Creeping Fingerwort
Leucobryum spp.: Large White-moss
Loeskeobryum brevirostre: Short-beaked Wood-moss
Lophocolea bidentata: Bifid Crestwort
Metzgeria furcata: Forked Veilwort
Mnium hornum: Swan's-neck Thyme-moss
Pellia endiviifolia: Endive Pellia
Pellia epiphylla: Overleaf Pellia
Plagiochila asplenioides: Greater Featherwort
Plagiochila porelloides: Lesser Featherwort
Plagiochila spinulosa: Prickly Featherwort
Plagiomnium undulatum: Hart's-tongue Thyme-moss
Plagiothecium undulatum: Waved Silk-moss
Pleurozia purpurea: Purple Spoonwort
Pleurozium schreberi: Red-stemmed Feather-moss
Pogonatum aloides: Aloe Haircap
Pogonatum urnigerum: Urn Haircap
Polytrichastrum alpinum: Alpine Haircap
Polytrichum commune: Common Haircap
Polytrichum formosum: Bank Haircap
Polytrichum juniperinum: Juniper Haircap
Polytrichum piliferum: Bristly Haircap
Polytrichum strictum: Strict Haircap
Pseudoscleropodium purum: Neat Feather-moss
Ptilium cristacastrensis: Ostrich-plume Feather-moss
Racomitrium lanuginosum: Woolly Fringe-moss
Rhizomnium punctatum: Dotted Thyme-moss
Rhytidiadelphus loreus: Little Shaggy-moss
Rhytidiadelphus squarrosus: Springy Turf-moss
Saccogyna viticulosa: Straggling Pouchwort
Scapania gracilis: Western Earwort
Scapania nemorea: Grove Earwort
Sphagnum auriculatum: Cow-horn Bog-moss
Sphagnum capillifolium: Red Bog-moss
Sphagnum compactum: Compact Bog-moss
Sphagnum contortum: Twisted Bog-moss
Sphagnum cuspidatum: Feathery Bog-moss
Sphagnum fallax: Flat-topped Bog-moss
Sphagnum fimbriatum: Fringed Bog-moss
Sphagnum girgensohnii: Girgensohn's Bog-moss
Sphagnum inundatum: Lesser Cow-horn Bog-moss
Sphagnum medium/divinum: Magellanic Bog-moss
Sphagnum palustre: Blunt-leaved Bog-moss
Sphagnum papillosum: Papillose Bog-moss
Sphagnum quinquefarium: Five-ranked Bog-moss
Sphagnum squarrosum: Spiky Bog-moss
Sphagnum strictum: Pale Bog-moss
Sphagnum subnitens: Lustrous Bog-moss
Sphagnum tenellum: Soft Bog-moss
Thamnobryum alopecurum: Fox-tail Feather-moss
Thuidium tamariscinum: Common Tamarisk-moss

Identification

Use page 304 to determine if your plant is a moss, liverwort, filmy-fern or clubmoss. If it is a filmy-fern it will be described on page 335. If it is a clubmoss, turn to page 336 on which all of our species are described except for four rarities. If it is a liverwort, check the common species described on pages 332–335.

If your plant is a moss, first look carefully to see if it is **unbranched** with each shoot being just a single upright main stem or **branched** with distinct side branches growing off the main stems. (Note that some unbranched mosses may have a few stems that divide into two.) Look also at the length of the leaves. In the table below, the combination of growth form (unbranched or branched) and leaf length (<3 mm or > 3 mm) will direct you to the appropriate pages.

<table>
<tr><th colspan="2">MAIN GROUPS OF MOSS SPECIES</th><th>Leaves up to 3 mm long</th><th>Leaves more than 3 mm long</th></tr>
<tr><td colspan="2" rowspan="2">Shoots consist mainly or entirely of upright, more or less UNBRANCHED STEMS with no side branches </td><td>Page 309 (some leaves of some of these species can be <3 mm long)</td><td>All species on pages 309–314</td></tr>
<tr><td colspan="2">Plagiothecium undulatum (p. 315) can fit here, in left or right columns as its leaves are mostly about 3 mm long</td></tr>
<tr><td rowspan="3">Shoots have MAIN STEMS AND SIDE BRANCHES</td><td rowspan="2">Side branches arise singly; not in groups at intervals along stem </td><td>Pages 315–319, except for Racomitrium lanuginosum (page 316) and Hylocomiadelphus triquetrus page 319)</td><td>Plagiomnium undulatum (page 314), Racomitrium lanuginosum (page 316) and Hylocomiadelphus triquetrus page 319)</td></tr>
<tr><td colspan="2">Note: Most leaves of Plagiothecium undulatum (page 315) and Rhytidiadelphus loreus (page 319) are about 3 mm long</td></tr>
<tr><td>Side branches in groups of 3 or more at intervals along upright main stem </td><td colspan="2">Sphagnum species (page 327)
Note: Another distinctive feature of Sphagnum mosses is the dense, flower-like cluster of young branches at the stem tip. These flower-like heads are what we see when looking down at a patch of Sphagnum moss.</td></tr>
</table>

Drawings to show growth forms of mosses and liverworts

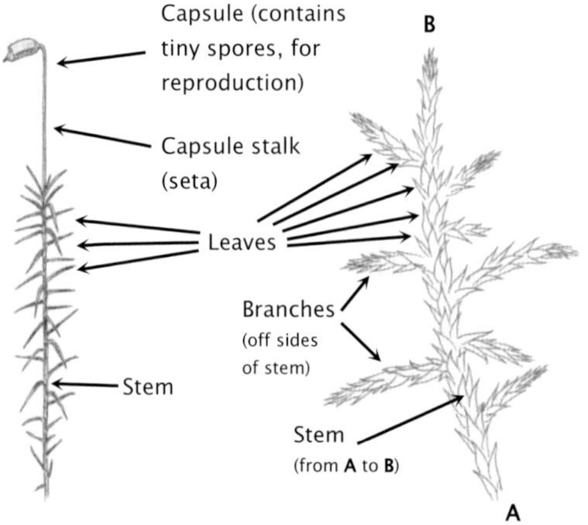

Unbranched moss

Branched moss

Moss leaf orientation

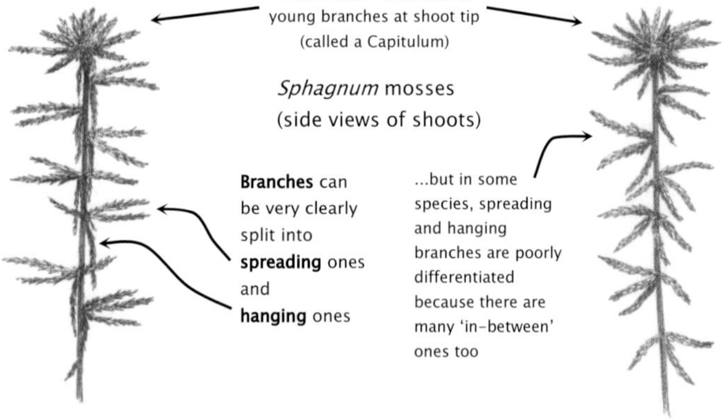

Leafy liverwort with round leaves

Leafy liverwort with lobed leaves

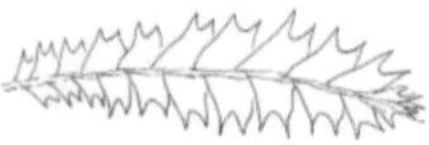

Thalloid liverwort

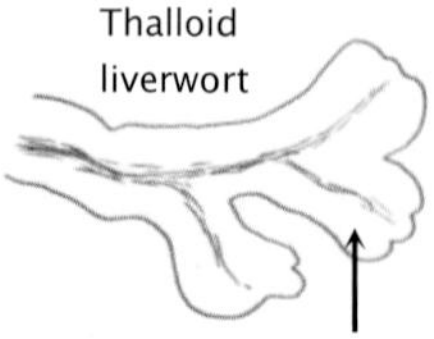

***Polytrichum* and related species.** Distinctive mosses with **stiff, opaque leaves** like little conifer needles **sticking out in all directions** (but pointing up when dry), on **upright, unbranched stems**. Each leaf has a translucent basal sheath held against the stem, and a longer blade sticking out. Capsules are oval or rectangular and on long upright stalks. Compare with *Dicranum* (page 311) and *Atrichum undulatum* (p. 314).

Capsule

Stem + leaves

More photos on next page >>

IDENTIFICATION

	Height (cm)	Leaf blade length (mm)	Leaf blade margin	Leaf tip	Capsule section	Other features	Leaf blade cross-section
Polytrichum commune	Up to 40 cm	8-12 mm	Toothed	Green	4-angled	Leaf sheath longer than wide	Rounded below
Polytrichum formosum	5-10 cm	8-12 mm	Toothed	Green	5-6-angled	Leaf sheath squarish	Ridged below
Polytrichastrum alpinum	4-10 cm	6-7 mm	Toothed	Green	not angled	Stem commonly divided into 2 or more branches	
Polytrichum piliferum	2-4 cm	3 mm	Un-toothed	White	4-angled	White 'hair-tip' can be quite long (1 mm)	
Polytrichum juniperinum	3-4 cm	3-5 mm	Un-toothed	Red-brown	4-angled	Red-brown at tip is very short	
Polytrichum strictum	Up to 40 cm	3-7 mm	Un-toothed	Red-brown	4-angled	White 'fuzzy stuff' (tomentum) on lower stem	
Pogonatum urnigerum	3 cm	5-10 mm	Toothed	Green	not angled	Leaf grey-green & relatively wide	
Pogonatum aloides	0.2-2 cm	3-4 mm	Toothed	Green	not angled	Leaf short but relatively wide	

HABITATS

Polytrichum commune	**Damp-wet acid mires** (M4-6), **woods** (W4) & **grasslands** (U4-6).
Polytrichum formosum	**Drier acid** soil in **woods** (W11/15-18) & **heaths**.
Polytrichastrum alpinum	**Damp or dry acid soils** in **upland heaths** & **mountain habitats**.
Polytrichum piliferum	**Dry, thin, acid soils** and **stony/gravelly/rocky ground** in **open places**, esp. where vegetation cover is only sparse.
Polytrichum juniperinum	
Polytrichum strictum	**Acid** mires, esp. **bogs** (M17-20).
Pogonatum urnigerum	**Acidic stony/gravelly/rocky ground** in **open places**.
Pogonatum aloides	**Acid soil**, esp. on **steep banks**.

Polytrichum and related species (2nd page)

Top: *Polytrichum commune.* Centre: *Polytrichum strictum.*
Bottom-left: *Polytrichum piliferum.* Bottom-right: *Pogonatum urnigerum.*

Dicranum scoparium has stems up to 10 cm tall, with **long (4-8 mm), narrow, bright green leaves, some of them curved to one side**. Forms cushions on **acid to neutral soil, rock and bark** in **woods, heaths, bogs, grasslands, dunes, cliffs and mountain habitats**. Paler & softer-textured than *Polytrichum* etc (page 309).

Dicranum majus ▶ is like a big *D. scoparium* up to 15 cm tall with **very long (9-15 mm) leaves very distinctly curved to one side**. Forms patches on **dry to moist, acid soils** in **upland Oak, Birch & Pine woods** (W11 & W17-18). Paler & softer-textured, with more curved leaves than *Polytrichum* etc (page 309).

Mnium hornum forms dense patches of unbranched stems 2–5 cm tall, with **dull green, narrow oval leaves** 4 mm long and spreading outwards, especially when moist. Young leaves in spring are pale. On **dry to moist, acid soil, rock and tree bases**, especially in **woods** (W4/10–11/14–18) but also in more open rocky places.

Rhizomnium punctatum ► has large **roundish leaves** up to 8 mm long on unbranched upright stems up to 10 cm tall. On **moist or wet, neutral to basic soils** and **among stones and rocks**, in **woods** (esp. W3, W5 & W7), **mires** (esp. M5, M9–11, M23, M32 & M37) and on **cliffs** (e.g. U17).

Leaves as rounded as in some liverworts, but with central 'vein' and not in two rows.

Campylopus atrovirens forms large, dense, **blackish cushions** 2-12 cm tall; **long narrow pointed leaves** up to 1 cm long, mostly with **white, hair-like extreme tips**. It grows on **moist or wet, acid peat** in **wet heaths** (M15) and **bogs** (M17), and on **flushed acid rocks** in **upland areas** in the **west**. Leaf shape as in *Dicranum* (page 311) but leaf colour different (much darker, and with white hair-point).

Leucobryum albidum / glaucum / juniperoideum ▸ has unbranched stems up to 20 cm tall, massed together into **dense whitish-green cushions** that can be very large. The **leaves are long (5-10 mm) and narrow**. Grows on **dry to moist, acid soils** in **woods** of Oak, Birch and Pine (mainly W11 & W17-18), and on damp or wet **heaths** (esp. M15-16). The three species are best separated from each other by features seen only under the microscope.

These two species have long, narrow **translucent leaves** with 'up & down' wave-like **undulations** running across them.

Atrichum undulatum ▸ has upright unbranched shoots mostly 2-6 cm tall, with **wavy leaves** up to 9 mm long sticking out in all directions and tapering to **pointed tips**. On dry to moist, **acid to neutral soils & stony ground** in **woods** (esp. W7-11), on **banks**, by **gravelly tracks** and **by rivers & streams**.

Plagiomnium undulatum has branched or unbranched shoots up to 10 cm tall, with **wavy leaves** that are up to 12 mm long and have **blunt or shortly-pointed tips**. On **acid, neutral and basic soils**, mainly in **woods** (esp. W7-12), **scrub** (W21-22) and **mires/fens** (esp. M22-23 & M27-28).

Plagiothecium undulatum is unmistakeable with its **flattened, whitish-green, sparingly branched shoots** up to 10 cm long, and **broad oval leaves** 3 mm long with **wavy undulations**. On moist **acid soils** in **woods** (W11 & W16-18 & **conifer plantations**), **heaths** (esp. H12, 16 & 21-22) & **bogs** (esp. M19). Shade-tolerant.

Ptilium cristacastrensis ▸ is equally unmistakeable, looking like patches of yellow-green feathers. The upright, **stiff, thick stems** are up to 10 cm tall, with many short, slender, **very closely and regularly spaced branches** whose **leaves are tiny (2-2.5 mm long), pointed and strongly curved to one side**. Shoot tips are thick and hooked. On **acid soils** in **upland woods** of Oak/Birch (W11 & 17) & Pine (W18); also in heaths (esp. H21). Mainly in the Scottish Highlands.

Racomitrium lanuginosum. A **large** moss whose branched shoots grow up to 15 cm long. Narrow, pointed leaves 4-6 mm long end in **white, hair-like extreme tips**, giving the whole plant a **grey tinge**. **Can form big grey-green patches** or hummocks. On **acid soil and rock** in **upland areas**: in **heaths** (esp. H10/14/15/17-21 & M15), **bogs** (M17-19), **grasslands** (U4-5, U7 & CG11), **cliffs**, **screes**, and **exposed mountain tops & ridges** (U9-10) where it can form extensive carpets (U10; **most luxuriant where grazing is light** - otherwise vegetation becomes more grassy).

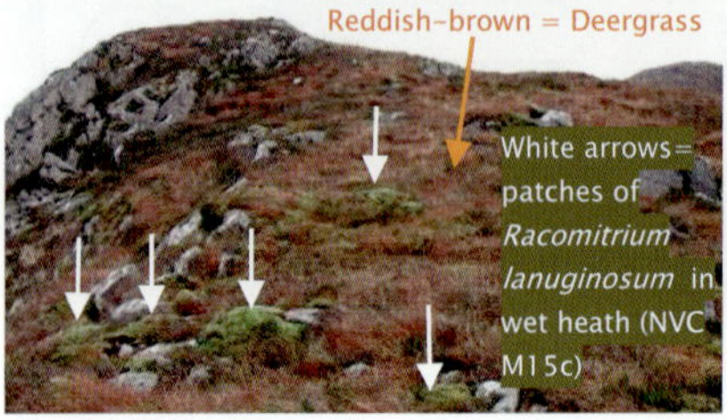

***Hypnum* species** are mosses with **small, pointed, regularly overlapping leaves curved to one side**. Shoots are slender, 2-5 cm long and branched.

H. andoi ► is very similar but **shoots tend to be narrower and commonly hang downwards** in thin mats; **leaves are slightly more curled** to one side; unopened **capsule** (inset) has **extremely short point** at tip. Commonest on **tree trunks and acid rocks in woods**, esp. in the west.

H. jutlandicum ► is more **regularly branched** and **paler green**, with leaves 1-2 mm long. Stems and branches can be rather flattened. It grows on **acid soils** in **heaths** and **bogs** (M17-20). Also on **acid rock** and **bark** in **woods** (esp. W17-18). Capsule less common than in *H. cupressiforme* var. *cup.* and *H. andoi.*

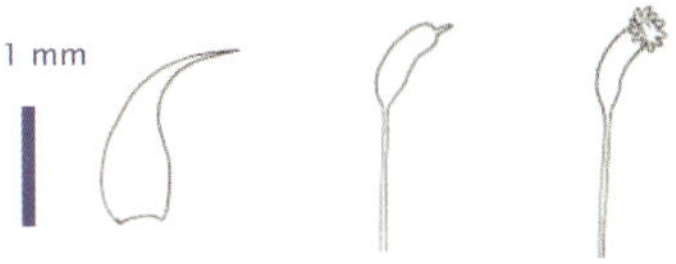

Leaf Capsules: unopened (L); opened (R)

◄ ***H. cupressiforme*** **var.** ***cupressiforme*** is **irregularly branched**, with thin branches and small leaves 1-2 mm long.Unopened capsule (inset) has short point at tip (but longer than in *H. andoi*). On **acid to basic rock and bark** in many habitats including **woodland** and **scrub**; also on **hedgerow trees, logs, tree stumps** and **stone walls**, and in **gardens**.

◄ ***H. cupressiforme*** **var.** ***lacunosum*** has **thicker, glossier, golden**, irregularly branched shoots with leaves 2-3 mm long, and grows mainly on **neutral to basic soils in grasslands** (CG1-2 & 7-14) and **sand dunes** (esp. SD7-12), and on **rocks** and **walls**. Capsule less common than in the last two taxa.

Hypnum can resemble *Brachythecium* (page 320), *Kindbergia* (page 322), *Homalothecium* (page 324) and *Isothecium* (page 325) but differs in its leaves being curled to one side.

Rhytidiadelphus squarrosus has upright, branched, **red stems** mostly 4-10 cm tall. The triangular, pointed **leaves** are 2-3 mm long and **bent outwards and backwards** giving star pattern when viewed from above. Leaves on side branches are straighter. Forms patches, mainly on **moist, acid, neutral or slightly basic soils**; in **grasslands** (esp. U4-6, MG5 & CG10-11), among **bracken** (U20), in **rush/sedge mires** (esp. M6 & M23), at **woodland edges** and in **garden lawns**.

The widely spreading, backward-pointing leaves are very distinctive. The related *R. loreus* (page 319) has a similar branching pattern as well as red stems and finely pointed leaves, but is larger, its stems are thicker and stiff, and its leaves are a little longer and are mostly curled to one side of the shoot.

Rhytidiadelphus loreus ► is a large, patch-forming moss with **stiff, red stems** up to 20 cm long, and side branches of varying length sticking out in various directions to give the plant a bushy look. Stems and branches bear **shiny leaves 3 mm long, with long narrow pointed tips curved mainly to one side**. On **dryish to moist, acid soils** in **woods** (esp. W11, 17 & 18), **grass-lands** (esp. U4-7, U13 & CG11), **heaths** (esp. H12, 16, 18 & 20-22) and **bogs** (esp. M19).

Compare with *R. squarrosus* (p. 318), *Hylocomiadelphus triquetrus* (below) & *Loeskeobryum brevirostre* (p. 321)

◄ ***Hylocomiadelphus triquetrus*** is like *R. loreus* (see above) but with **paler, non-shiny, triangular leaves up to 6 mm long, with broad bases**. The large, wide-based leaves give the shoots a thick, bushy, bulky look. On **dryish to moist, neutral to basic and (esp. in NE) acid soils**, mainly in **woods** (esp. W9, 11, 17 & 18) but also in **sand dunes** (esp. SD7-8 & 12).

Brachythecium rutabulum is **irregularly branched**, has rather thick, soft-textured, **pale-ended shoots** up to 12 cm long, and **pale green leaves 2-3 mm long, tapering to fine points**. The leaves are neither widely spreading nor held tightly against the shoots. It grows on **moist neutral soils** in **woods** (W1-10), **scrub** (W21/22), **grasslands** (esp. MG1), **mires** (esp. M22/23 & 27/28), **hedgerows** & **gardens**.

B. rivulare is very similar and not always separable without microscopic examination, though it tends to have **more regularly spaced and more widely-spreading branches**, and its **leaves are held closer to the shoots**. It is less common and grows mainly on **wet neutral soils** in **mires/flushes** (e.g. M23 & M32) and **woods** (W1-3 & 5-7).

Common species that are rather similar: *Eurhynchium striatum* (page 321) is stiffer, with shorter, triangular, 'pleated' leaves. *Pseudoscleropodium purum* has more regular branching, and its leaves have blunter ends and 'sudden' short, curved extreme tips. *Kindbergia praelonga* (page 322), with which *B. rutabulum* grows very commonly, has narrower, more closely and more regularly spaced branches with smaller leaves, so it is much more delicate-looking than *B. rutabulum*.

Eurhynchium striatum has **green stems** (5–10 cm long) and **triangular leaves** up to 2 mm long. It is common on dry to slightly moist **neutral to basic soils** in **woods** (mainly W8–9 but also some W10–11) in both **upland and lowland areas.**

Loeskeobryum brevirostre ► is similar but with **red stems**. Oval **leaf suddenly narrowed near pointed tip**. Less common than *E. striatum*; mainly in **upland woods** of Ash, Hazel, Rowan, oak and birch (W9 & W11). Compare also with *Rhytidiadelphus loreus* (page 319).

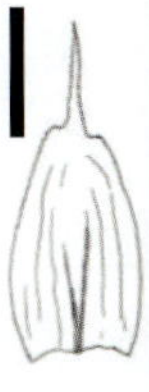

Kindbergia praelonga has green stems 1-12 cm long. **Branches are thin and delicate** with **tiny, narrow oval, finely pointed leaves** 1-1.5 mm long. Leaves on stems are larger (1-2 mm) and oval-triangular with pointed tips. Some branches have a few smaller secondary branches. *Thuidium tamariscinum* (page 323) is a similar size and colour but has more intricate branching, smaller, blunter branch leaves, stiffer stems and branches, and a matt texture. *K. praelonga* grows mostly on **moist neutral soils** in **woods** (esp. W1-10), **scrub** (W21-22), **mires** (esp. M22-23 & 27-28), **hedgerows** and **gardens**.

Hylocomium splendens is large, with **red stems** up to 20 cm long. **Branches intricately subdivided**. Branch leaves (R inset) tiny, 1 mm long; stem leaves (L inset) larger, 2-3 mm. Leaves **rather glossy**. Mostly on **acid soils** in upland **woods** (W11 & 17-18), **heaths** (H10, 12, 16 & 18-22), **bogs** (esp. M19) and **grasslands** (U4-8/13). Also in upland calcareous grasslands (CG10-12) & *Dryas* heath (CG13-14).

Thuidium tamariscinum is similar but has **green stems** (older, lower parts can be browner or blackish) and **even more intricately divided branches** with tiny leaves 0.5 mm long. Stem leaves larger (1-2 mm). Leaves not glossy; plant has a **matt** finish. On **dry to moist, acid to basic rock, soil, tree bases** & **logs** in **woods** (W4, 7, 9, 11, 17-18), **scrub** (W19), **bracken** (W25, U20) and some **heaths** (H12 & 21).

Homalothecium sericeum has **stems** up to about 10 cm long, **creeping** on rock or bark and bearing **many short, closely-spaced branches** that are **<1 cm long and curl upwards and inwards when dry**. The leaves are 2-3 mm long, **glossy golden-green**, very **narrow** and tapering to **long fine points**. Through a lens one can see pleat-like folds running the length of the leaf, giving the leaf a **striped** look. This species grows on **dry, neutral to basic rocks, walls and bark** (e.g. Ash, Elm, Sycamore and Elder). It is commonest in the lowlands.

Homalothecium sericeum (moist)

Homalothecium sericeum (dry)

Homalothecium lutescens (moist)

Homalothecium lutescens is similar to *H. sericeum* but the main shoots are **more upright** and the **branches are longer (>1 cm)** and **not so curled when dry**. It also differs in that it **grows on soil** rather than rock or bark. It is found in **dry basic unimproved grassland**, especially CG1-9 in the lowlands. It also grows in **sand dunes** (esp. SD7-8).

Both species can resemble *Hypnum* (page 317 - but *Hypnum* leaves curve to one side of stem) and *Brachythecium* (page 320 - but the *Brachythecium* species have wider leaves).

Isothecium myosuroides has small, finely-pointed leaves 1–2 mm long. Massed shoots can form big, dull (matt; not shiny) olive-green patches. It can look rather like *Hypnum* (page 317) but is more bushy and matt-textured, and its leaves are straight and not curved to one side of stem. It grows on acid rock and bark (esp. oak, birch & alder) in woods (esp. W11 & 17–18) & ravines. ***I. alopecuroides*** is similarly bushy and dull-coloured but its thicker branches have blunter, concave leaves; on neutral to basic rock and bark in woods (esp. W7–11) and by streams.

Thamnobryum alopecurum is larger, with shoots 5–15 cm long, their upper parts with many closely spaced branches (some subdivided) spreading out to right and left as in a fern frond. The shoots are a dull dark green colour and form patches that can be very large. It grows mainly on steep, shaded, moist to wet, neutral to basic rocks by streams, especially in woods.

These three mosses have side branches spaced quite regularly along left and right sides of the stems, and their **leaves are blunt-tipped** (with a tiny projecting point in *Pseudoscleropodium*). Stem reddish in the first two species; green in the third.

Pleurozium schreberi ▸ has **red stems** up to 12 cm long, with **rather curved side branches**. Leaves 2-3 mm long. On **acid soils** in **woods** (esp. W17-18), in **heaths** (esp. H10, 12, 16 & 18-22), **bogs** (M17-20), and in **grasslands** (U2-7 & 13).

◂ ***Calliergonella cuspidata*** has **red stems** (but can be green in shade) up to 12 cm long, with many **straight branches.** Leaves 2-3 mm long. Young leaves folded into **sharp points at stem and branch tips**. On **damp or wet, neutral to basic soils** in **woods** (esp. W1-3 & 5-7), **mires** (esp. M5, 8-13 & 22-28), **springs** (M32 & 37-38) and **dune slacks** (SD14-17).

Pseudoscleropodium purum ▸ has **green stems**, branches pointed or blunt-ended, and **leaf 2 mm long with tiny curved projection at tip**. On **dry to moist, acid to basic soil** in **woods** (esp. W10-11 & 17-18), **grasslands** (U4, CG10), **heaths** (H10/12) and among patches of **bracken** (U20/W25).

***Sphagnum* species** have branches (some spreading; some hanging) in groups at intervals along the main stem. Youngest branches form flower-like clusters at stem tips. Summary table below; descriptions and photos on following four pages.

	**General colour**	**Stem colour**	**Branch width** (thin <1mm med.1-2mm thick >2mm) **and shape**	**Branc-hes per group**	**Spreading & hanging branches clearly diff-erentiated?**	**Branch leaf length** (mm)
palustre	green to orange-pink	green or brown	thick; straightish; tapering	4-6	Yes	1.7-2.8
papillosum	greenish to yellow-brown	green or brown	thick; straightish; blunt	3-4	Yes	1.5-2.5
medium / divinum	dark reddish	red	thick; straightish; blunt	3-5	Yes	1.7-2.7
squarrosum	green	green or brown	thick; straightish; 'prickly' look	4-6	Yes	1.7-3.1
compactum	green, yellow orange	dark brown/ black	medium-thickness; straightish	2-7	Yes (but branches v. dense)	1.8-3.0
strictum	pale green or orange	pale greenish	med.; straightish	3-6	Yes (but v. dense)	1.7-2.7
auriculatum	green, yell-ow, orange, red-brown	dark brown/ black	medium to thick; can be curved	3-5	No	1.3-2.5
inundatum	mostly greenish to yellowish	dark brown/ black	medium-thickn.; can be curved	4-6	Yes	1.3-2.2
contortum	green, yellowish, or brownish	pale brown/ green	medium-thickness; curved	4-6	Yes	1.0-1.9
subnitens	green, brown, pink; glossy when dry	brown	medium to thin; straightish	3	Yes	1.2-2.7
capillifolium	red or mixed green & red	red or green	thin; straightish	3-4	Yes	0.8-1.4
quinquefarium	green or green + pink	green	thin; straightish	4-5	Yes	0.8-1.6
fallax	green to orange	green to orange	thin; straightish	5	Yes	1.2-2.1
fimbriatum	green	green	thin; straightish	3-4	Yes	0.8-2.0
girgensohnii	green	green	thin; straightish	3-4	Yes	1.1-1.6
cuspidatum	green	green	thin; straight or curved	3-5	No	1.5-3.5
tenellum	green, yellow orange	green	thin; straightish	3	No	0.8-1.5

Sphagnum palustre ► is **large** and green to pale orange-brown, with **thick branches tapering to pointed tips**. Branch leaves have tips folded in like 'hoods'. Shoot tip has distinctly raised centre that can be a different (richer and less green) colour than the surrounding branches. Common in **acid to neutral mires** (M4-6) and **wet woods** (W3-4), but scarce in bogs.

◄ ***S. papillosum*** is **large**, like *S. palustre*, but mostly **ochre**-coloured, with most **branches short and blunt-tipped**, and whole shoot tip all one colour and with centre not markedly raised. Mainly in wet acid **bogs** (M17/18/21). Scarce in wet heath (M15), rush/sedge mire (M4/6 & U6) and damp acid grassland (U5).

S. medium/divinum ► like *S. palustre* and *S. papillosum* in size and shape, but **dull red**. Centre of shoot tip does not project above level of the surrounding branches and is the same colour as the surrounding branches. More or less confined to **bogs**, especially wetter examples of M18 vegetation on raised bogs.

◄ ***S. squarrosum*** is **green** and looks like *S. palustre* but the **branch leaves bend outwards** so the **branches look 'prickly'**. Unlike most *Sphagnum* species it grows mainly on wet, **neutral to basic soils**. It is found in **woods** (esp. W3) and **sedge/herb mires** (esp. M5).

Sphagnum compactum ► forms **low, dense, yellow, green or orange patches**. The shoots have **dark stems** and **very crowded, densely-packed branches**. Most **branches at shoot tips point upwards**; in most other *Sphagnum* species many upper branches stick out sideways. On **moist, acid peat** in **wet heaths** (M15-16) and **bogs** (M17-18). Commonest in wet heaths that have been **burned**.

S. strictum is similar but has paler, more washed-out colours, **pale stems** and **sticking-out branch leaves** ('prickly' like *S. squarrosum* but plant much denser). **Acid peat** in **wet heaths** (M15) & **bogs** (M17); W Scotland & W Ireland (+ very rare in N Wales).

Sphagnum compactum

Sphagnum auriculatum

◄ ***S. auriculatum*** is **golden, green, yellow or brown** with a **dark stem**. Many branches are **curved** like cow's horns; **some hang down, some stick out, and others are in-between**. Branch leaves don't stick out much from branches. In **acid & neutral mires** (M4-12), **wet heaths** (M15-16) & **springs** (M31-32). Mainly where there is **lateral groundwater movement** but also in some **bog pools** (M1).

S. inundatum is similar in looks and habitats, but **hanging and spreading branches are distinct from each other**, and branch leaves stick out a little from the branches.

S. contortum is like *S. inundatum* but paler and with a **pale stem**. It grows in **neutral to basic mires** and is less common than the last two species.

Sphagnum subnitens ► Mostly pinkish or brownish. More flesh-pink than in *S. capillifolium/rubellum*. Can have a greener centre to the shoot tip. Branches longer and thicker (1-1.5 mm) than in *S. capillifolium / rubellum*. In a wide range of acid to neutral mires & fens and also in bogs and wet woods.

◄ ***S. capillifolium/rubellum*** is our commonest 'Red *Sphagnum*'. Thin branches up to c. 1 mm wide, in groups of 3-4. Greener in shade but some red always present. Main stem mostly red. In acid mires (M4, M6-7), damp or wet heaths (M15 & H21-22), bogs (M17-20) and damp or wet woods (W4 & W17-18).

S. capillifolium: dense, domed shoot tip; stem leaf >1.2 mm long with a pointed tip. ***S. rubellum***: looser, flatter shoot tip; stem leaf <1.2 mm long, with a blunter tip.

S. quinquefarium ► is like *S. capillifolium/rubellum* but greener and dotted/flecked with pink, has a green stem and denser branches in groups of 4-5. In acid Oak-Birch-Pine woods (W17-18) and damp heaths (H21) in upland areas (N & W); esp. on N- to E-facing slopes.

Sphagnum fallax ► is **mostly green but can be yellowish or orange**, has a rather weak and floppy **pale stem**, and thin branches. Spreading and hanging branches are very distinct from each other. Most leaves on main stems point downwards (upwards in *S. fimbriatum* & *S. girgensohnii* - see below). Common in **acidic mires** (esp. M2-4 & M6-7), **bogs** (esp. M18 & M20-21) and **wet woods** (W3-4).

S. fimbriatum

S. cuspidatum ► is also like *S. fallax* and is **pale green or yellow-green**, but has **distinctively long, narrow branch leaves** up to 3.5 mm long, giving it a soft and 'wispy' look. It grows mainly in **very wet, acidic depressions** among **bogs**: especially in **bog pools** (M1 & M2) where it **can be aquatic**, but also scattered in the extensive main bog vegetation (M17 & M18).

◄ ***S. fimbriatum*** and ***S. girgensohnii*** are rather similar; **green**, with **leaves on the main stems pointing up** and with **very wide, jaggy tips in *S. fimbriatum*** and **narrower jaggy tips in *S. girgensohnii***; if you pull the top off a shoot, stem leaves stick up prominently from the break-point. On **wet, acid ground** in **woods** (esp. W4) and **mires** (esp. M6).

Submerged

◄ ***S. tenellum*** is **small, yellow- or orange-green** with **short, thin branches** mostly <5 mm long. Hanging & spreading branches not clearly separated. **Branch leaves are short and oval, and can diverge from each other at branch tips**. On **wet acid peat** in **bogs** (M17-18) & **wet heaths** (M15-16); can be common where previously **burned**.

Lophocolea bidentata ► has **flattened shoots** with two rows of **translucent pale green leaves**. Each leaf ends in **two sharply-pointed lobes**. On **moist, acid, neutral or basic soils**, and on **rocks, logs and tree bases**. Mainly in **shaded or sheltered** places, especially **woods**; also in **mires**, **heaths** and **grasslands**.

◄ ***Diplophyllum albicans*** has **flattened shoots** with two rows of **dull green leaves** that are **oblong and blunt** with a **pale line up the centre**, and, pressed close against their uppersides, what look like **similar but smaller** leaves (actually **lobes** - each leaf has big & small lobes). Mainly on **acid rock & soil** in places with some **shade or shelter**, especially **woods**.

Plagiochila asplenioides ► is **large**, with **pale** green shoots up to 12 cm long and 9 mm wide. Its **roundish leaves** have tiny teeth on their edges. Commonest on **moist, neutral to basic soils** in **woods** (esp. W7-9). ***P. porelloides*** is **similar but smaller** (shoots <6 mm wide) and **darker**; common on **acid to basic rocks, walls & soil** in sheltered places such as **woods** (esp. W8-11 & 17) & **by streams**.

◄ ***Lepidozia reptans*** is **small**, with **regularly and closely branched, pale green shoots** creeping over surfaces of **rock, soil, logs or bark**. Mainly in **woods**. Appearance of stems and side branches quite distinctive (as shown in photo). Leaves are tiny and cut deeply into 3-4 narrow, finger-like lobes (lens needed to see these).

Frullania **species** ► have branched shoots forming **distinct purple-brown patches**. Leaves tiny and round, with complex-shaped additional lobes and 'underleaves' (visible with a lens). Five species (two of them common) on **well-drained rock and bark surfaces**. Commonest in **woods** but also in more open places (e.g. on soil in short grassland, and on coastal rocks).

◄ ***Metzgeria furcata*** has a **branched, narrow, parallel-sided pale or dull green, flattened, creeping thallus** looking like roads on a map. Forms **flat patches** on well-drained **neutral to basic rock and bark** (e.g. Ash, Sycamore, Hazel and Willows); **mainly in woods**.

M. consanguinea/violacea: similar but with **narrow, outward-projecting thalli tips**. Mainly on **bark**. *M. violacea* widespread; *M. consanguinea* mainly on acid bark in western areas. Separation requires close look at positions of pale powdery gemmae on thalli tips.

Conocephalum conicum ► is **large**, with a **shiny green flattened thallus with small pale dots on its surface** (not smooth as in *Pellia*). Forms **conspicuous patches** - that can be very large - on **moist, neutral to basic soily and rocky banks** in **woods** (esp. W7-9).

Pellia endiviifolia

P. neesiana: like *P. epiphylla* but **aromatic** (+ v. small-scale differences). In similar habitats, esp. flushes.

◄ ***Pellia*** **species:**

P. epiphylla has a simple, **smooth, rounded, wavy-edged thallus** with a broad, slightly darker or red- or purple-tinted line up the middle. Common on **moist or wet, acid to neutral soil or rock**, especially in **woods**, in **mires** and on steep banks by **streams** and **ditches**.

P. endiviifolia is similar but thallus edges grow **frilly branches** in autumn (old decayed ones remaining into following year). No red tint. It grows on **moist or wet, basic rock & soil**, especially in woods.

A few western woodland liverworts

The species on this page are common and conspicuous in many rocky, sheltered woods in W Scotland, W Ireland, the Lake District & Wales - areas of international importance for these and other liverworts. Colours vary from green to brownish.

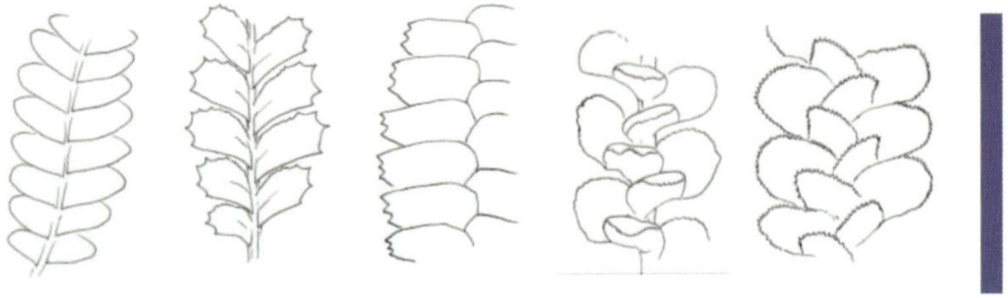

Saccogyna viticulosa *Plagiochila spinulosa* *Bazzania trilobata* *Scapania gracilis* *Scapania nemorea* 1 cm

Saccogyna viticulosa has untoothed, unlobed **leaves in opposite pairs**. Creeping shoots form pale green to brown patches on **steep, moist, acid to basic rocks and banks** in **sheltered, humid western woods** (W9/11/17).

Plagiochila spinulosa has **large teeth along the wide rounded leaf tips**. Forms greenish or brownish patches on **rocks** and **trees** in **sheltered, humid western woods** (W9/11/17-18); less commonly in sheltered **heaths** (H21) and on **cliffs**.

Bazzania trilobata is **large, bright green and conspicuous**. Its **rectangular leaves have three small teeth along their otherwise straight, square-cut tips**. Forms patches on **acid rocks, banks** and **tree bases** in **western woods** (W11/17-18).

Scapania gracilis has **leaves with two, unequally-sized, rounded lobes. Small lobe bends out away from large lobe. Each lobe has small, short, inconspicuous marginal teeth.** Greenish-brownish patches on **acid rocks** & **trees** in **western woods** (W9/11/17-18); also in western **heaths** (esp. H20-21). (***S. nemorea*** is similar but widespread (west + east) on acid-neutral rock/soil/logs in woods; small lobe pressed against big one; marginal teeth longer, sharper and more numerous.)

A distinctive western liverwort of more open places...

Pleurozia purpurea ► has **purplish leaves that overlap in such way that the upright shoots are worm-like in shape**. Forms distinctive patches on **moist or wet, acid peat** in **heaths** (H20-21 and M15) and **bogs** (mainly M17) in W Scotland & W Ireland.

Filmy-ferns *Hymenophyllum* species

To most people these plants look like large, dark green liverworts, which is why they are described here. Their **evergreen shoots** are up to 10 cm long and 3 cm wide and are **translucent dark green**. Massed shoots form **patches** on steep **acid rocks, banks and tree trunks** in **sheltered, humid, undisturbed habitats** in the **west**.

Wilson's Filmy-fern *H. wilsonii* has shoots **sticking out** from rock/bank/bark, with their **side lobes** (pinnae) **sparingly branched** and **twisted at an angle** to give the whole shoot some **depth** when viewed side-on. In **ancient woods** (W9/11/17), **heaths** (H20-21) and **rocky habitats** in the west, from SW England northwards.

Tunbridge Filmy-fern *H. tunbrigense* has **bluer-green** shoots that mostly **hang down** against rock/bank/bark, and **well-branched side lobes all held in the same plane** so the **shoot is flatter**. On **steep, rocky banks** in **ancient woods** (W11/17) in west (N to Skye/Raasay) and (rare) in SE England. Scarcer than Wilson's Filmy-fern.

Wilson's Filmy-fern

Tunbridge Filmy-fern

Alpine Clubmoss ► *Diphasiastrum alpinum* has **creeping** shoots with **short, neatly overlapping leaves that are 2–4 mm long and don't stick out**. In **short grassland or heath** on **thin or stony acid soils** in **exposed places** in the **hills**, especially at high altitudes. Can be abundant where the vegetation is kept short by **grazing** (by sheep and/or deer).

◄ **Fir Clubmoss** *Huperzia selago* has unmistakeable **tufts of upright shoots** mostly <20 cm tall, with **spreading, narrow, pointed leaves 4–8 mm long**. In **short or open grassland or heath** on **thin or stony acid soils** in the **hills**, especially at high altitudes. Most common where the vegetation is short by **grazing** (by sheep and/or deer).

Stag's-horn Clubmoss ► *Lycopodium clavatum* has **creeping** shoots and **narrow leaves 3–5 mm long.** Has a **white, very thin, hair-like extreme leaf tip**. In **grassland or heath** on **thin acid to basic soils** and on **stony ground**, especially in upland areas. Mainly where vegetation is kept shorter or open by **grazing** (by sheep and/or deer) or by **disturbance** (e.g. gravelly tracksides).

Interrupted Clubmoss *L. annotinum* is like Stag's-horn Clubmoss but its **leaves are slightly larger (4–6 mm long) and lack white hair-like tips**. Uncommon in **damp, acid, heathery heaths** (esp. H21–22) in the **Scottish Highlands**; very rare in the Lake District.

Lesser Clubmoss *Selaginella selaginoides* is **small**, with creeping and upright shoots <5 cm tall and **leaves 2–4 mm long, lacking hair-points but with toothed edges**. Mainly in **base-rich flushes** (esp. M10–12) and **damp grasslands** (esp. CG9–11) in upland areas.

A few lichens

Lichens and fungi do not belong in the Plant Kingdom, but a few of our 1700 species of lichens are described briefly below. Lichens are extraordinary 'combination' organisms composed mainly of fungus material that provides a home for smaller algae (or in some species bacteria) that help to 'feed' the fungus. This symbiotic (you help me, I help you) relationship allows lichens to occupy a wider range of habitats than would be possible if each separate fungus and algal or bacterial component had to 'go it alone'.

Cladonia portentosa, C. ciliata & C. arbuscula ▸ **Creamy** coloured **and intricately branched**. Branch tips stick out in all directions in *C. portentosa* and bend to one side in *C. ciliata & C. arbuscula*. Extreme branch tips divide into 2 very tiny points in *C. ciliata* and 3–5 tiny points in *C. arbuscula*. *C. arbuscula* has thicker branches than the other two. All are common on **dry to damp, acid soils**, in **heaths, bogs & grasslands**. *C. arbuscula* prefers damp ground.

◂ ***Cladonia uncialis***. **Creamy** coloured, with **thick, pointed branches**. Not intricately branched as in the last three species. Common on **dry to damp, acid soils**, in **heaths, bogs and grasslands**.

Usnea **species** ▸ **Pale grey or green-grey and finely branched** as in *Cladonia portentosa/ciliata/arbuscula* but **less stiff and commonly hanging down**. Grows on **acid bark** of many tree and shrub species, and on **acid rock**. Commonest in **woods** in **clean air** parts of the north and west.

◂ ***Hypogymnia physodes***. **Small and pale grey, with intricate lobes with black undersides**. Common on **acid bark** of many types of tree and shrub, and also on dwarf shrubs such as Heather and Bilberry. Also on **acid rock**. Mainly in **woods** and **heaths**.

Lobaria pulmonaria ▸ **Large, olive-green and unmistakeable with ridges on upper surface**. Mainly in **woods** in N & W, on **neutral to basic bark of Ash, Elm & Hazel** but also on Oak, Willow & other trees, and on rocks. Uncommon outside W Highlands, but a good, conspicuous indicator of **clean air** and **limited disturbance to woodland**.

HABITATS

This section complements the rest of the book in taking a wider view of habitats and vegetation – an illustrated tour through the various habitats and vegetation types of Britain and Ireland, ordered by broad habitat type. It describes the main vegetation types and how they relate to each other, to soil types and to land management.

National Vegetation Classification (NVC) codes are given wherever appropriate in this section. The NVC is the standard British vegetation classification system and is used for surveys, conservation assessments and so on. The NVC is based on mixtures (communities) of plant species growing together. For example, vegetation made up mainly of Heather + Bell Heather = NVC community H10; Heather + Sand Sedge = H11; Heather + Bilberry = H12. The study of plant communities adds greatly to anyone's understanding and appreciation of landscapes at all scales and from scientific and aesthetic viewpoints. Full names of NVC communities are given in a list on pages 378–380, followed by lists of habitat categories (and NVC relationships) for a few other habitat classifications and notes on yet more classification systems. The NVC is a British classification but is quite usable in Ireland too, though there are also two separate classifications of Irish vegetation (see page 384).

The photo on this page is labelled with NVC codes to show how vegetation types can vary within a small area. A key to these NVC codes is given below:

M6 *Carex echinata-Sphagnum fallax/auriculatum* mire (acid mire with Soft Rush and *Sphagnum* mosses). **M15** *Trichophorum germanicum-Erica tetralix* wet heath (wet heath with Deergrass, Purple Moor-grass, Heather and Cross-leaved Heath). **M17** *Trichophorum germanicum-Eriophorum vaginatum* bog (with Heather, Cross-leaved Heath, Deergrass, Purple Moor-grass, Common & Hare's-tail Cottongrasses and *Sphagnum* mosses, with hummocks of moss *Racomitrium lanuginosum*, one of these labelled ***Rac. lan.***). **M25** *Molinia caerulea-Potentilla erecta* mire (dominated by dense tussocks of Purple Moor-grass). **U4** *Festuca ovina-Agrostis capillaris-Galium saxatile* grassland (with Common Bent, Sheep's Fescue, Tormentil, Heath Bedstraw and mosses including *Hylocomium splendens* & *Hypnum jutlandicum*). **U5** *Nardus stricta-Galium saxatile* grassland (dominated by Mat Grass).

An example of a habitat-mapped landscape is shown on the next page.

An example of vegetation classification on a landscape scale (using NVC)

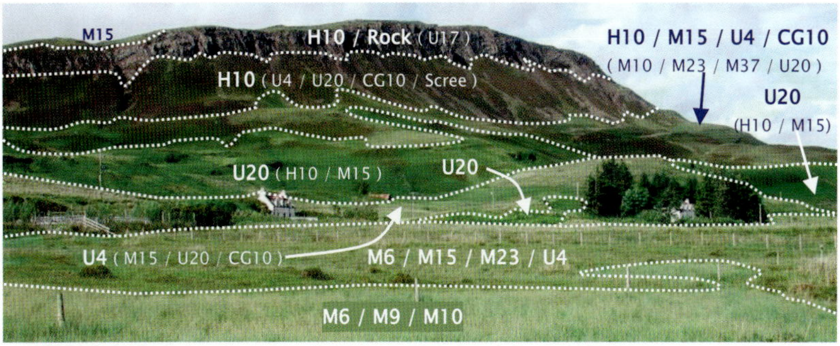

Simplified key: H10 dry heath; M6 acid sedge mire; M9 & M10 basic sedge mires; M15 wet heath; M23 neutral rush mire; M37 calcareous spring; U4 acid grassland; U17 tall herb veg.; U20 bracken; CG10 calcareous grassland. This example shows mapping simplified for the purpose of the photo, amalgamating mapped units into fewer, larger polygons. In the mosaic labels bracketed codes indicate minor components. An estimated percentage figure for each NVC code is current normal practice for mosaic labels but would lead to visual complication in a photo-illustration such as this.

Definition of 'Upland' and 'Lowland' as used in this book

The terms 'upland' and 'lowland' as used in this book and by many ecologists are defined by climate more than by altitude. **Upland** environments have a colder, wetter climate, a greater extent of bog and wet heath, and a greater abundance of mosses, liverworts and ferns. They have a north-western distribution and extend down to sea level in the far north and west. Most upland areas have a predominance of old, hard rocks and agriculturally poor soils. **Lowland** environments are warmer and drier and have a southern and eastern distribution. Their rocks are generally younger and softer, and their soils more agriculturally productive, than in upland areas. The differences in vegetation between these two zones are reflected in the NVC, which has separate north-western (upland) and south-eastern (lowland) plant communities within several broad habitat groupings such as woods, heaths, acid grasslands and calcareous grasslands.

Map of climatic severity in Britain and Ireland

This map is based on an index of climatic severity calculated as mean annual precipitation (cm) divided by annual mean temperature (°C), based on data for 1981-2010 from Blockeel, T.L., Bosanquet, S.D.S., Hill, M.O. & Preston, C.D.; (2014); *Atlas of British and Irish Bryophytes*; 2 vols; Pisces Publications, https://www.metoffice.gov.uk/ and https://www.met.ie/climate-ireland/SummaryClimAvgs.pdf

Colours indicate index values: white = <12; blue = 12-24; grey = 25-49; brown = 50+. Higher values = a more severe (colder/wetter) climate. Geographical variation in index values is reflected in that of vegetation species composition. A threshold of 12 could be interpreted as a threshold between **upland (cooler/wetter)** and **lowland (warmer/drier)** climates and environments. Other indices, based on slightly different measures such as rainfall frequency or annual accumulated temperature give broadly similar geographical distribution patterns.

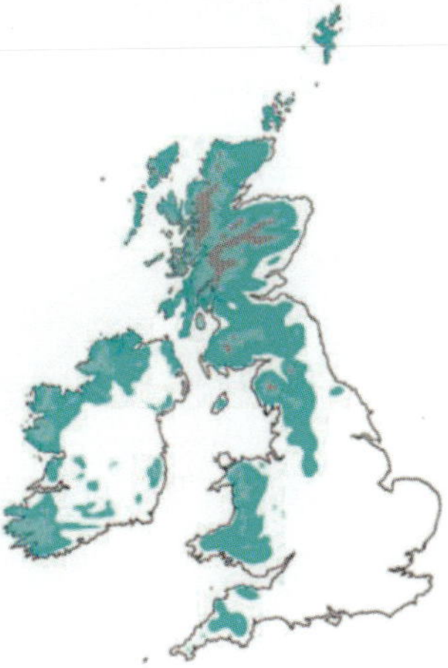

Woodland and scrub

Our native woodland vegetation varies greatly in relation to the wetness and acidity of the soil, hence the following broad divisions into dry woods and wet woods, each of which is subdivided according to the acidity of the soils.

Woods on **dry (or, at least, well-drained) acid** soils mostly have canopies of birches, oaks (W16/17), Scots Pine (W18) or Beech (W15), and ground vegetation including various combinations of Heather, Bilberry (these dwarf shrubs especially common where grazing is only light), Wavy Hair-grass and mosses (grass/moss tending to dominate where more heavily grazed).

Lush Bilberry in lightly-grazed birch woodland (NVC W17)

Lightly-grazed pinewood (W18) with tall Heather

Moss carpets in heavily-grazed oak woodland (NVC W17)

Heavily-grazed Beechwood with Wavy Hair-grass and sparse moss growth (NVC W15)

Woods on **dry/well-drained neutral** (or slightly acid) soils typically have varied mixtures of trees and shrubs including oaks, birches, Rowan, Holly, Hazel (W10/11) Beech (W14), Sycamore, Sweet Chestnut, Hornbeam and Elder (W10), and a ground flora including Yorkshire-fog, Creeping Soft-grass, Sweet vernal-grass, Wood Sorrel, Wood Anemone, Bluebell, Bracken and, where grazing is light, Bramble, Raspberry, Honeysuckle and Buckler-ferns.

Above + right: mixed deciduous lowland woodland with grassy ground layer (NVC W10) in winter; lightly grazed

Oak woodland with Bluebells (NVC W10) in spring; medium-grazed

Tormentil, Heath Bedstraw, Common Dog-violet and mosses including *Hylocomiadelphus triquetrus*, in upland birchwood (NVC W11)

Woods on **dry/well-drained basic** soils typically have canopies including Ash, Wych Elm, Sycamore, Hazel, limes (W8/9), Rowan, Downy Birch (W9), Field Maple, Elder (W8), Beech (W12) and Yew (W13), and a ground flora including at least some of Dog's Mercury, Sanicle, Woodruff, Hedge Woundwort, Herb-Robert, Wood Avens, Ground Ivy, Bluebell, False Brome and Hairy-brome.

Dog's Mercury and Wood Sorrel beneath Ash, Sycamore and Elder (NVC W9); lightly to moderately grazed

Ash, Dog's Mercury, Wild Garlic, ferns and mosses (NVC W9); lightly to moderately grazed

Dog's Mercury, Red Campion, Wild Garlic, Nettle, Goosegrass and Primrose beneath Ash and Sycamore (NVC W8); lightly-grazed

Wet woods generally have canopies of willows, birches and Alder. Those on **wet acid** soils (W4) commonly have a ground flora including Purple Moor-grass and mosses including *Polytrichum commune* and *Sphagnum* species. Those on **wet neutral to basic soils** (W1-3 & W5-7) typically have more herb-rich ground floras including species such as Meadowsweet, Water Avens, Marsh Bedstraw, Stinging Nettle, Goosegrass, Yellow Flag, Opposite-leaved Golden Saxifrage, Tufted Hair-grass and Reed Canary-grass.

Purple Moor-grass beneath Downy Birch (NVC W4); medium-grazed

Yellow Flag, Marsh Hawksbeard and other herbs, beneath Alders (NVC W7); lightly to moderately grazed

Tufted Hair-grass beneath Alders (NVC W7); lightly grazed (hence tall grass swards)

Native woodland also varies in relation to climate. Through much of Britain, especially in the lowlands, the various types of woodland vegetation have much in common with those found in **temperate broadleaved forest** in western and central Europe. In some western areas with a mild and very wet climate (W Highlands, W Ireland and smaller areas in Galloway, the Lake District, Wales and SW England) there is such a luxuriance and diversity of mosses, liverworts, lichens and ferns that the woodland is **temperate rainforest** with similarities to rainforests in places such as W Canada, New Zealand, the Himalayan foothills and many tropical mountains. In the colder and drier north-east of Scotland the combination of birch, pine, Juniper and plants such as Chickweed-wintergreen is similar to that in Scandinavian **boreal forests**.

Temperate rainforest in the Western Scottish Highlands (NVC W17)

Boreal birch/juniper woodland in NE Scotland (NVC W17/19)

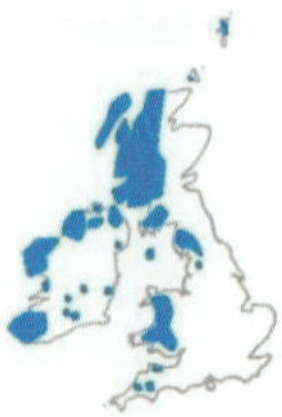

◄ In Britain and Ireland temperate rainforest is found mainly in the blue zone, where mean annual precipitation (cm) / maximum deviation of monthly mean temperatures from 9ºC gives a 'climatic temperate rainforest-ness' index value of >25 (wet + not too hot + not too cold).

Temperate broadleaved woodland in lowland southern Scotland (NVC W10)

Scrub is like woodland but with a canopy of bushes such as Hawthorn (W21), Blackthorn (W22), Gorse/Broom (W23), Elder (W21), Juniper (W19), Sea-buckthorn (SD19), Hazel (scrub forms of W9 & W11), and willows (W1, scrub forms of W2-7 and, on some mountains, rare willows forming W20).

Tall neutral grassland (NVC MG1) just downslope of scrub. Could develop into Hawthorn/Elder scrub (W21) if grazing is sufficiently light.

Blackthorn scrub (NVC W22) on neutral-basic soils in lowlands. Has potential to develop into taller woodland with Ash and Sycamore (W8), esp. if grazing is very light.

Gorse scrub (NVC W23)

Gorse scrub (NVC W23) on neutral soils in the lowlands. Has potential here to develop into taller lowland woodland (W8/10). Gorse scrub in upland areas can develop into other types of woodland, especially W11/17 incl. oak, birch, Goat Willow and Rowan.

In Juniper scrub the bushes are mainly old; juniper regeneration is scarce.

Juniper scrub (NVC W19) on acid soils in uplands of E Scotland, typically with some bushes upright and others spreading

Sea Buckthorn scrub (NVC SD18) on sand dunes. Hawthorn and Elder can colonize (W21); then Ash and Sycamore (W8).

Woodland Management

Woodland once covered most of Britain and Ireland. Over many centuries people cleared trees to make farmland. They brought in sheep and cattle, increasing the numbers of large herbivores (which include our native red and roe deer) that browse tree seedlings and saplings and thereby prevent many young trees from growing. Felling, grazing/browsing, and, in upland areas, climatic deterioration (cool, wet climate through most of the last 10,000 years producing leached, infertile soils less well-suited to tree growth) have reduced the area of native woodland so that we now have just a small fraction of the original extent.

Grazing

Lightly grazed or ungrazed woodland has tall ground vegetation (whose plants are able to flower freely) and can develop a dense shrub layer. More heavily grazed woodland has shorter, more grassy or mossy ground vegetation and, if heavily grazed for a long time, little or no shrub layer, a canopy dominated by mature trees and a scarcity or absence of young trees. Light grazing is often seen as beneficial as regards the flowering of grasses and herbs and the development of a varied tree/shrub layer including young trees. However, it can eventually lead to tall, species-poor ground vegetation and a general increase in shade from the dense tree/shrub layer; this can all get too shaded for some scarce woodland plants and lichens that need some shade but not too much, so it is good to try to strike a balance and get some of the benefits of lighter grazing but not in such a way as to compromise other aspects of the woodland vegetation.

Heavily grazed oakwood with no shrub layer

Lightly grazed oakwood with young Holly

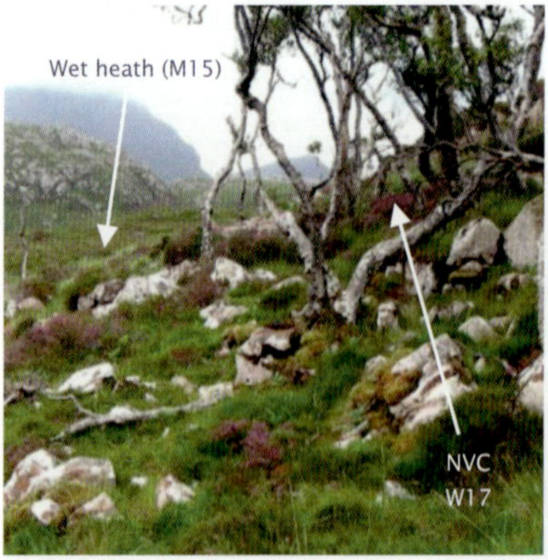

Heavily grazed birchwood with no young trees

Ungrazed deer-fenced natural regeneration exclosure with tall vegetation & young trees

Coppicing

Many woods have in the past been coppiced: tree/shrub stems cut almost at ground level so that over the next 10–20 years several thin, straightish stems grows from the cut base. When they are long enough to be used for various purposes such as fencing, basket-making and charcoal production, they are cut and the cycle begins again. For a few years after coppicing, conditions are more open and well lit: this can encourage the growth and flowering of many woodland herbs, resulting in a 'flush' of colour during the spring and summer months of these years. This is another reason why some woods are coppiced nowadays, for nature conservation purposes. However, in the north-west, and especially in the western Highlands and Hebrides, where many uncommon shade-demanding mosses, liverworts and lichens grow on woodland trees and shrubs, coppicing can be damaging because it removes the habitat of these small plants and lets in too much light. Some moss, liverwort and lichen communities such as those on older, thicker trunks and branches take a long time to develop, and coppicing can understandably limit or prevent their development.

Hazel 14 years after coppicing

Hazel with no sign of coppicing and with rich flora of mosses, liverworts and lichens

Multi-stemmed oak, coppiced long ago. Ground vegetation quite heavily grazed, leading to dominance of mosses and lack of dwarf shrubs that might otherwise grow on these acid soils.

Ancient Woods

Most - or perhaps all - of the remaining woodland has probably been modified in some way, directly or indirectly, by human activities. The amount of modification varies. Many woods appear relatively undisturbed and have been present for at least a few centuries: these are termed Ancient Woods. Ancient Woodland indicator species grow best where there has been long historical continuity of shade and shelter beneath a relatively undisturbed tree canopy. They include Soft Shield-fern, Hay-scented Buckler-fern, Wood Sedge, Remote Sedge, Hairy-brome, Giant Fescue, Sanicle, Woodruff, Enchanter's Nightshade, Dog's Mercury, Wood Speedwell, Wild Garlic and Bluebell (noting that in the west, Bluebell also grows outside woods, mainly among Bracken but also in grassland and on sea cliffs). The more of these species are in a wood, the more likely it is that the wood is ancient. In southern Britain more species can be indicators: for example, Greater & Hairy Woodrushes, Wood Anemone and Wood Sorrel; further N & W they also grow outside woods because the cooler, wetter climate compensates for the lack of shade.

Dog's Mercury

NVC W8

Planted forest and natural regeneration 'new native woodland' projects

Planted forests vary from dense stands of non-native species such as Sitka Spruce and Lodgepole Pine (common and extensive on acid soils in upland areas) to more open plantations of native or non-native conifers and broadleaves. Where more light gets in, the ground vegetation is generally better developed and, in many places, can be classified to a NVC community. Some planted forests contain open glades with interesting heaths, grasslands and wetlands; in Scotland the Forestry Commission (now Forestry & Land Scotland) is actively managing a number of these sites to benefit key woodland species and surveying the remainder to decide on conservation-related management, including deer management, controlled grazing and the removal of non-native species.

Upland conifer plantation in Perthshire, with blocks of trees of different ages

Lowland Sycamore/Pine plantation with well-developed ground vegetation

NVC W8

Recent decades have seen many woodland creation projects aimed at increasing the cover of native trees and shrubs. Some are plantations and others utilize natural regeneration; either way, grazing is generally kept minimal for many years to allow the young trees to reach a good size. These projects can benefit conservation, especially where previous ecological interest was low (e.g. agriculturally improved or very unnaturally species-poor grassland, or dense bracken). In other places they can be a problem, for example by breaking up extensive upland heath/bog landscapes of high ecological interest, or because some common, vigorous plants (especially grasses) can, when ungrazed or only very lightly grazed, grow so tall and dense as to smother smaller plants of interest, especially as deep mats of dead grass litter can accumulate over the years in ungrazed or minimally grazed places. Some short, grazed vegetation with an abundance of small plants is important botanically or in other ways such as for insects. Many woodland regeneration projects therefore aim for a good balance between tree regeneration and other kinds of ecological interest; this balance can be attempted in various ways such as getting the right level of grazing, or removing grazing for only a limited period of time. Not surprisingly, habitat surveys are commonly carried out in advance of these new woodland projects.

Very pale old dead leaves of tall ungrazed Purple Moor-grass (NVC M25 + M15)

Deer-fenced exclosure to allow native woodland to expand out from remnant in stream gully. Photographed in spring, nine years after the fence was put up.

Young planted Scots Pine and birch in heath (NVC H12 & M15), deer-fenced to exclude grazing

Invasive non-native species

Some non-native plant species can colonize woodland and form such extensive, dense, overshading thickets that most native plants on the ground, rocks and lower tree trunks can no longer survive. Among the best known of these invaders, especially in the densely-populated lowlands, are Japanese Knotweed and Himalayan Balsam. However, Rhododendron is particularly damaging because it chokes our western temperate rainforests whose rich floras of mosses, liverworts and lichens make them among the most internationally important habitats for nature conservation in Britain and Ireland. Much work is being done to remove Rhododendron from these places.

Various methods of Rhododendron removal are in use, including cut & burn, stem injection and lever & mulch. The last two can be very effective. Lever & mulch was devised by Gordon French and Donald Kennedy in the western Scottish Highlands and is described below (description based on information provided by Donald; see also https://savingscotlandsrainforest.org.uk/people/contractor):

The bushes are dismantled or uprooted by human force, with some strategic use of saws and hammer. The aim is to remove all bud-bearing wood of all stems that have root connections. Any buds on stems that cannot be removed are bashed with a hammer. A mulch of leaf-litter and the cut brash is then covered over the stump or 'node', to reduce the chance of regrowth and recovery. On revisits (ideally 3 within 5 years, and one about 10 years after initial treatment) any missed or incompletely killed material should be easy to spot and finish off with hammer, saw, adze or bare hands.

A few years after Lever & Mulch Rhododendron eradication. The brash will eventually decay.

Other invasive species in British and Irish native woodland include various conifers (commonly seeding out from nearby plantations), Beech (which can cause serious overshading and is therefore often a problem, at least outside its restricted southern native range in Britain), Sycamore (which can grow quite dense and cast plenty of shade, though in many north-western areas its nutrient-rich bark provides habitat for many lichens of special interest).

Hedgerows, Underscrub and Tall Ferns

Hedgerows ► have narrow, linear strips of modified woodland/scrub (Hawthorn especially common), and other habitats such as bramble underscrub (W24), grassland (esp. MG1) and weedy vegetation (esp. OV24-27). Typical plants include Garlic Mustard, Cow Parsley, Hedge Parsley and Rough Chervil. In lowland farmland hedgerows are valuable 'corridors' of 'wilder' habitat. Older hedgerows generally have a greater diversity of tree & shrub species.

Bramble (NVC W24)

◄ **Underscrub** consists mainly of tall (up to 1.5 m), dense, ungrazed or lightly grazed Bramble (W24) or other species such as Raspberry. It is found mainly in sheltered places at low altitudes: in woodland glades, along woodland margins, along hedgerows and roadsides, around the edges of fields and on disturbed ground.

Bracken ► can grow so tall and dense as to dominate the vegetation, either within woodland (esp. in W10-11 & W14-17) or in the open (U20 & W25). Bracken patches are mostly on deep, acid to neutral soils and can be extensive, commonly in association with grasslands (esp. U4) and heaths (various types). Bracken encroachment can cause losses of ecological interest and agricultural productivity. Measures to prevent or limit its spread include cutting, rolling, chemical treatment and trampling by cattle.

NVC U20

Scaly Male Fern

◄ **Other tall ferns** are common in woods (esp. W10-11 & W14-17) and also form patches on acid to neutral soils on open hillsides: these patches on open ground are mainly of Lemon-scented Fern (U19) and Scaly Male Fern (no NVC type).

Grasslands

These consist mainly of grasses, though some grass-dominated vegetation is classed as wetland where the dominant grasses are Purple Moor-grass (M24-26), Common Reed (S4 & S24-26), Reed Canary-grass (S28, Floating Sweet-grass (S22) and Reed Sweet-grass (S5) (the 'S' NVC codes are swamps/fens). The main divisions within grasslands are based on the acidity of the soils.

Acid grasslands (U1-7) have swards of bent Grasses, fescues, Sweet Vernal-grass, Wavy Hair-grass, Mat Grass and Heath Rush, with herbs such as Heath Bedstraw, Tormentil and Sheep's Sorrel. They can be very mossy. They are common and extensive on acid soils in grazed uplands but also occur more patchily in the lowlands. Soft Rush is common in some acid grasslands.

Fescue/bent acid grassland with Tormentil & Heath Bedstraw (NVC U4). Maintained by heavy grazing; otherwise it would be taller, less herb-rich and perhaps more heathy.

Neutral grasslands have varied swards of grasses such as False Oat-grass, Cocksfoot, Crested Dog's-tail & Tufted Hair-grass. They are not very mossy. Some are rushy (Soft or Hard Rushes). Some (esp. MG2-5) have rich mixtures of herbs. Soils vary from quite dry (MG1/3/5) to damp (MG2/4/8-16). Management is typically pasture grazed by cattle, sheep or horses, or meadow mown in summer for hay/silage. They are found mainly at low altitudes.

Unimproved neutral grassland with Crested Dog's-tail, Red Fescue, Devil's-bit Scabious, Knapweed, Ribwort Plantain, Yarrow and Autumn Hawkbit (NVC MG5). Medium-grazed. If grazing is very light (or absent) the vegetation will become taller and less herb-rich: more like MG1 (below left).

Tall, lightly grazed, species-poor sward of False Oat-grass (NVC MG1). Has potential to become scrub (W21) and broadleaved woodland (esp. W8/10) .

Tall, lightly-grazed, species-poor sward of Tufted Hair-grass (NVC MG9). Has potential to become wet alder / willow / birch woodland (W7).

These two tall, lightly grazed neutral grasslands can become scrub & woodland if grazing continues to be no more than very light for many years.

Calcareous grasslands (CG1–14) commonly include bent grasses, fescues, Meadow Oat-grass, Quaking Grass, Glaucous Sedge, Thyme (the single most characteristic indicator species of calcareous grasslands), Rockrose and Common Bird's-foot-trefoil. Can be mossy. Not rushy. Mostly short and grazed, but some in the south (CG3–6) have tall, little-grazed swards including Tor-grass or Upright Brome. Found in uplands and lowlands, on thin basic soils that are mostly dry but in some places damp and flushed.

Unimproved, sheep-grazed upland calcareous grassland with Sheep's Fescue, Common Bent, Sweet Vernal-grass and Thyme (NVC CG10); grazing intensity medium to high.

Extensive grazed fescue-bent upland grasslands with Thyme (NVC CG10/11) and, on deeper, more acid soils, Heath Bedstraw (U4)

Tall, little-grazed sward of Tor-grass (NVC CG4) in southern England

Grassland management

Agriculturally-improved Rye-grass pasture (NVC MG7)

Agriculturally-improved grasslands (MG6-7) have been treated with fertilisers or herbicides, or reseeded, and have bright green species-poor swards of mainly Perennial Rye-grass, White Clover, Rough Meadow-grass and Timothy. Mainly in the lowlands, where they are very common and extensive.

Unimproved grasslands (U1-7, MG1-5/8-16 & CG1-14), as shown in photos on the last three pages have not been so heavily treated. Their richer floras typically include at least some of Mat Grass, Wavy Hair-grass, Sheep's Fescue, False Oat-grass, Purple Moor-grass, Tufted Hair-grass, Sweet Vernal-grass, Quaking Grass, Bent grasses, Heath Rush, Woodrushes, Sedges, Bulbous Buttercup, Wood Crane's-bill, Tormentil, Common Bird's-foot-trefoil, Red Clover, Heath & Lady's Bedstraws, Selfheal, Devil's-bit Scabious, Thyme, Rockrose, Yellow-rattle, Ribwort Plantain, Primrose, Cowslip, Knapweeds, Mouse-ear Hawkweed, Violets and Orchids. Mosses can be abundant. **Heavily grazed** grasslands are short: many plants have their flower heads bitten off but small herbs (and Thyme & Rockrose) can be abundant. **Lightly grazed** grasslands are taller: plants flower well but small ones can be outcompeted by tall plants or smothered by mats of dead grass leaves that build up over the years. **Meadows** are mown at least one a year, so they go through tall and short phases and can contain a great abundance and diversity of plant species.

Herb-rich neutral grassland (NVC MG5) that has been only lightly grazed in recent years

Thick mat of dead grass leaves in very lightly grazed species-poor neutral grassland (NVC MG1)

Medium-grazed herb-rich neutral grassland (NVC MG5)

Heaths

Heath is vegetation in which dwarf shrubs such as Heather, Bell Heather and Bilberry are abundant or dominant. **Wet heath** (M14–16, H3, 5, 6 & some H4 & H7) is on moist or wet peat and includes Cross-leaved Heath, Purple Moor-grass, Deergrass or Bog-myrtle. **Dry heath** (H1, 2, 8–22 & some H4 & H7) differs in that those four species are rare or absent and the vegetation is mostly dominated by Heather or Bilberry (Bilberry commonest where a cold climate or heavy grazing restricts Heather growth). Wet heath can look like bog but has little or no Hare's-tail Cottongrass, *Sphagnum papillosum* or *S. medium/divinum* and it extends onto thinner soil and steeper slopes. Gorse (page 44) is common in many low altitude heaths, growing with Western Gorse in Ireland and S. Britain and with Dwarf Gorse in S England.

Dry heath with Heather and Bell Heather

Wet heath with Heather, Cross-leaved Heath, Purple Moor-grass, Deergrass and Bog-myrtle

Note: *Some people have defined heath as heathery vegetation on lowland mineral soil, contrasted with 'moor' on upland peat, but this misses out heathy vegetation on lowland peat or upland mineral soil, and to add confusion the term 'moor' is often used for landscape-scale mixtures of heath, bog and acid grassland. It is clearer to define heath as any upland or lowland dwarf shrub vegetation that is not bog: this is the approach taken by the NVC.*

Heaths are commonest in the north and west of Britain and Ireland. The extent of heaths and associated bogs in our uplands is important in a global context. Wet heath is most extensive in the west Highlands, the Hebrides and western Ireland. Dry heath is the dominant type further east.

Typical West Highland upland: a mosaic of heath (mainly wet heath - NVC M15) and blanket bog (M17/19). Upland Britain is of international significance for its extensive heath/bog moorlands. In autumn the colours of Deergrass and Purple Moor-grass turn these places into beautifully golden landscapes.

Montane Bilberry-Crowberry dry heath (NVC H18/20/22) on Schiehallion, in the Highlands. Bilberry/Crowberry dominance is natural here (too cold for Heather).

Montane prostrate Heather heath (NVC H14) on a windy ridge top in the Highlands. Frequent strong wind makes the Heather grow in a very low, creeping way. Britain & Ireland (esp. Scotland) is almost the only part of Europe with prostrate Heather heath.

Dry Heather-dominated heaths are common in many areas at low to medium altitude, mainly in upland areas of the N & W. Note the division between those containing Bell Heather (mainly on sunny S-W-facing slopes and commonest in the milder west) and those with Bilberry (on a wide range of slope aspects and commonest in colder places and in more eastern uplands).

Heather with Bell Heather (NVC H10)

Heather with Bilberry (NVC H12)

Heather + Bilberry + western liverworts (H21b): a rare heath of cool, wet areas in the NW.

Western Gorse, Heather and Bell Heather (NVC H8): a heath of the warmer S and W.

Heathland management

Dwarf shrubs generally grow taller and thicker in lightly grazed heaths than in more heavily grazed heaths (which are typically more grassy). Heather condition can give useful clues about past and present grazing: see pages 52–56. Heather, Bell Heather and Bilberry are palatable to deer, sheep and cattle, and are often grazed. Cross-leaved Heath, Cowberry, Crowberry and Bearberry are unpalatable and grazed much less; grazed plants of these species suggest a high herbivore density. Not all heaths contain abundant dwarf shrubs: some heavily grazed wet heath is dominated by Deergrass, with dwarf shrubs rare or even absent.

In many heaths Heather is burned periodically for commercial grouse-shooting (burning small patches to produce short & tall mosaics favourable to Red Grouse) or to rejuvenate heather to provide a sweeter bite for sheep. After burning, the vegetation regrows with a more even, uniform structure and a less diverse moss, liverwort and lichen flora. Abundant Juniper suggests a lack of intensive burning. For conservation management, burning is not recommended in heaths generally, especially those of high botanical interest (e.g. with Juniper, Arctic Bearberry, Bog Bilberry, Dwarf Birch or rich floras of mosses, liverworts and lichens), or those on very steep slopes (which can be botanically rich). Abundant Bearberry in heath that is not steep or rocky suggests previous burning: Bearberry can gain a stronger competitive edge after burning has reduced the height and density of Heather.

Patch-burning for grouse-moor management. Paler patches = most recently burned. See also photographs on page 57.

Wetlands

This is a very large and varied group of habitats, broken down here into four main sub-groups: bogs, aquatic habitats, swamps and other wetlands. It is convenient to look first at bogs because they are very distinct from other types of wetlands in terms of their vegetation and hydrology.

Bogs

A bog is a wetland on deep, wet, acidic peat that receives most of its water from rainfall. More water comes in as rain than leaves by means of evaporation or (through vegetation) transpiration. This is why bogs are restricted to cool, wet parts of the world such as Britain and Ireland. Bog vegetation (M17–21) is a mixture of grasses, sedges, dwarf shrubs and mosses, and includes good quantities of one or more of Hare's-tail Cottongrass and the mosses *Sphagnum papillosum* and *S. medium/divinum*. It varies from heathery (little-grazed) to grass/sedge-dominated (more heavily grazed). **Blanket bogs** (M17–20) cover large areas of level and gently sloping peat-covered ground in the N & W (mainly where mean annual precipitation is >1m and summer mean temperatures <14°C); their large extent is significant in a global context. **Raised bogs** (mainly M18) have convex, dome-like profiles and are mainly at low altitudes from Wales and the English Midlands northwards. **Valley bogs** (M21) are widespread but uncommon on deep, very wet peat in depressions and valley bottoms. The name is misleading because most bog in valleys is actually blanket or raised bog. The 'true' valley bog plant community (M21) is like very wet blanket bog but with some species that indicate a very little sluggish lateral water movement through the vegetation. Small **bog pools** (M1–3) with *Sphagnum* mosses and Common Cottongrass, and deeper pools with Bogbean (no NVC type) or just open water, are scattered among many bogs, especially in blanket bogs in N Scotland.

Vegetation typical of the main wetter bog plant communities (NVC M17 and M18)

The main types of bog are shown in these photographs:

Wet blanket bog with abundant *Sphagnum* mosses (NVC M17). Commonest at low to medium altitude in western upland areas. This example, patterned with bog pools, is in NW Sutherland.

Raised bog. Scattered at low altitudes from Wales northwards. Their main vegetation type (NVC M18, as shown here in the N Cumbria lowlands) is like M17 blanket bog but with more Cranberry, Bog-rosemary, *Sphagnum medium/divinum* and *S. fallax*, and less Tormentil.

Heather-Hare's-tail cottongrass blanket bog (NVC M19) at about 500 m in Perthshire. This kind of bog is slightly drier and more heathery than the wetter M17 and M18 types and is widespread on watersheds and gentle slopes, mainly at medium to high altitudes.

Valley bog (NVC M21) in a wet valley bottom in the W Highlands. Found mainly in SW Britain, this uncommon type of habitat is transitional between a bog and a very wet, sluggish flush.

See also Hare's-tail Cottongrass bog (M20); photo on next page.

Effects of land management on bogs

Draining or burning can make bogs drier, with more Heather and Purple Moor-grass and a reduced abundance and diversity of *Sphagnum* mosses and other bog plants. Heavy grazing can cause losses of dwarf shrubs and increases of grasses and sedges. Combinations of draining, burning and heavy grazing can cause bog vegetation to become wet heath (M15), Purple Moor-grass mire (M25), Hare's-tail Cottongrass bog (M20) or Heath Rush-dominated vegetation (U6). Trampling by deer and sheep can add to natural erosion to produce 'hagged' bog with mosaics of peat 'islands' and intervening wet channels or bare peat. If grazing is very light or absent, plants such as Heather can become so tall and dominant that they draw up more ground water and cause the habitat to become slightly drier with a reduced extent and diversity of *Sphagnum* mosses and other small plants. Trees such as birch and pine can colonize the drier parts of some bogs, especially on lowland raised bogs; this appears to be a natural process, perhaps speeded up where bogs have become drier as a result of draining or burning. On some drained or previously conifer-planted bogs, drains are now being blocked or dammed in an aim to raise the water table and maintain the wet bog habitat and vegetation, and ex-plantation ridge/furrow topography 'smoothed'. Commercial peat removal (e.g. for horticultural use) is detrimental to the ecological interest of bogs because it completely strips off the upper layer of peat, with all the vegetation.

Heavily grazed bog in W Lothian; Hare's-tail Cottongrass dominant; no dwarf shrubs (NVC M20). Heavy grazing can produce this type of vegetation on both raised and blanket bogs.

Hagged blanket bog (M17/19) with eroded channels and raised peat 'islands'; erosion evidently exacerbated by sheep trampling

Young birches and pines colonizing a lowland raised bog

Aquatic habitats

These are the wettest of all wetland habitats, with varying assemblages (NVC A1–24) of aquatic or 'emergent' plants growing in shallow to quite deep still or running water.

Loch with White Water-lilies

Bog pools with Bogbean

Swamps

These habitats (NVC S1–23) consist mainly of species-poor swards of tall sedges or grasses growing in shallow water or on extremely wet, more or less level ground, especially by lakes and rivers.

Swamps around a loch margin in southern Scotland.

Bulrush swamp

Note: *in other parts of the world the term swamp refers to woodland that is subjected to flooding by waters of adjacent rivers or lakes.*

Other wetlands

This very large and very varied group of habitats can be divided into the following sub-types, based mainly on the nature of the wetness:

Springs occur where water emerges from underground. They are mostly small and situated on hill slopes. Their vegetation (M31-38) is a mixture of small herbs, sedges, grasses, mosses and liverworts.

Flushes, also known as ***soligenous mires***, occur where water flows or seeps over the ground. Many are long and narrow, but they can be wider. They occur in many situations. It is common to find flushes just downslope of springs. Flush vegetation (including M4-14/15a/23-28 & MG8-10) includes various mixtures of rushes, sedges, herbs and mosses.

Basin mires, also called ***topogenous mires,*** form in the bottoms of wet hollows, with little or no lateral water movement. Their vegetation (including M1-6/9/23) consists mainly of rushes, sedges, herbs and mosses. Swamp, fen/marsh and bog vegetation types can also occur in basin situations.

Fen and ***Marsh*** are terms used mainly for wetlands on various wet, neutral to basic soils, with vegetation (including M9-10/13/22-28, S4/24-28 & MG8-13) typically dominated by tall grasses, sedges or rushes. Some can look like swamps but are less wet and have an extensive lower layer of herbs and mosses or mats of dead plant material. The terms fen and marsh have been used to indicate peat and mineral soils respectively, but there is overlap in the vegetation types on both soil types and also overlap with the vegetation and habitats of flushes, basin mires and inundation wetlands.

Fen with Common Reed and Meadowsweet (NVC S4). Ungrazed.

Meadowsweet fen (NVC M27). Lightly grazed.

Neutral wetland with Sharp-flowered Rush and Devil's-bit Scabious (NVC M23). Medium-grazed.

Soakways are like extremely wet, slow-moving flushes winding their way through other wetland habitats. Their vegetation (including M1/29–30 & M35) consists mainly of herbs, sedges, small rushes, and mosses.

Inundation habitats occur in places frequently flooded by fresh water: for example, around ponds and lakes, and by rivers. Their vegetation (M30, OV28–36 & MG13) is made up mainly of herbs and grasses. ***Dune slacks*** are a particular coastal kind of inundation habitat: wet, periodically flooded hollows among sand dunes, with vegetation (SD13–17) consisting mainly of herbs (including Silverweed), grasses, sedges, mosses and Creeping Willow. Wetlands inundated by sea water are very different: see Saltmarshes (p. 375).

Soakway with Bog Pondweed (NVC M30)

Dune slack with varied mix of plant communities including swamps

Acidic Purple Moor-grass mire (M25a/b), dominated by tussocks of Purple Moor-grass, can occur in flushes but is more extensive over large spreads of acidic, peaty soil in western Britain in habitats that do not fit well into any of the above-described types and whose vegetation (M25a/b) is probably derived from wet heath (M15/16) or bog (M17/19) as a result of burning and grazing.

Purple Moor-grass mire (NVC M25) on acid, peaty soil. Medium-grazed. If less grazed it would be more heathy (this example probably wet heath NVC M15).

The definitions of these sub-types, most of which are, together with bogs, within the much broader NVC grouping of ***mires***, may appear confusing because of their overlapping vegetation types and habitats, especially between fen/marsh, flush and basin mire. However, when looking at the vegetation types irrespective of flushing, inundation or whether the substrate is peat or mineral soil, they can be fairly readily put into three groups reflecting the acidity of the soils:

Acid (M1-4, M6-7, M24c, M25a, M31, M32a & M33), with sedges, rushes, dwarf shrubs and grasses; the lower layer of the vegetation including few herb species but abundant mosses including some of *Sphagnum palustre, S. fallax, S. auriculatum, S. capillifolium, S. rubellum* and *Polytrichum commune*).

Neutral (M5, M22-23, M24a/b, M25b/c, M26-30, M32b, M35-36, S24-28, SD15-17, OV28-36, MG8-13 & some M9), with the above mosses scarce or absent and replaced by species such as *Calliergonella cuspidata* and *Brachythecium rivulare*, herbs such as Meadowsweet, Angelica and Marsh Bedstraw, and sedges such as Glaucous, Carnation and Brown Sedges).

Basic (M8, M10-13, 34, 37-38, SD13-14 & some M9), with sedges including Dioecious Sedge, Flea Sedge, Broad-leaved Cottongrass or Black Bog-rush, and the lower layer including 'brown mosses' such as *Palustriella commutata*.

These photos show close views of examples of each of these three groups:

Acid
(with *Sphagnum/Polytrichum*)

Neutral
(with mesotrophic herbs)

Basic
(with 'brown mosses')

Wetland management

In wetlands generally, grazing produces shorter and more grass/sedge/rush-dominated vegetation. For example, it can convert Meadowsweet-dominated vegetation (M27) into rush mire (M23). Too much grazing and associated trampling can be damaging, as for example in flushes with a sparse flora of small plants. However, grazing can help to maintain good botanical diversity, and very light grazing or a lack of grazing can lead to over-dominance of common, vigorous species (that can take up more groundwater and have a drying-out effect on the whole habitat), losses of smaller plants and a reduction in botanical diversity.

Heavily grazed/trampled

Medium-grazed

Very lightly grazed

Nutrient enrichment (eutrophication) from sources such as agricultural fertilizer and urine from sheep or cattle can cause increases of plants such as Stinging Nettle and Goosegrass.

Yellow Flag mire with abundant Stinging Nettle indicating nutrient enrichment

Drying out of wetland habitats by artificial draining can affect the vegetation in various ways. For example, species that prefer slightly drier conditions can become more abundant, in some places leading to the development of scrub and woodland. Areas of wetland can become mosaics of wetter and drier types where criss-crossed by drains. Natural small-scale variation in wetness and groundwater behaviour, and associated irregular mosaics of different wetland vegetation types, can be altered or lost by artificial drainage. The planting of trees in wetland can lead to general drying out as the trees grow and take up more groundwater, as well as altering the habitat by casting shade and leaf litter.

Wetland with small-scale variation in hydrology, vegetation structure and species composition reflecting natural physical conditions, and with no clear indications of anthropogenic overgrazing, undergrazing, eutrophication or drainage.

Montane habitats

Some of the scrub, grassland, heath and wetland vegetation types already described occur on and among the higher hills of Britain and Ireland, but on much of the high ground where the climate is very harsh and the soils thin, stony and poor in nutrients there are specialized types of montane habitat and vegetation. These include plant species well adapted to the harsh environment. Some of these species can be grown successfully in well-weeded lowland gardens, suggesting that their restriction in the wild to montane habitats is at least partly because they are outcompeted elsewhere by more vigorous plants that cannot tolerate the harsh montane climate.

On the highest and most exposed summits and ridges, grasslands and dwarf shrub heaths give way (with increasing altitude and exposure) to montane **summit heaths** (U8–10), the commonest of which (U10) contains extensive carpets of the grey-green moss *Racomitrium lanuginosum* dotted with other plants including Stiff Sedge and Dwarf Willow (see below).

Racomitrium lanuginosum moss-heath on a high, exposed ridge

R. lanuginosum with **Dwarf Willow** *Salix herbacea*, which is like Bilberry (page 62) but very **short/creeping**, with **round to oval, net-veined leaves**; in **exposed montane heaths** (esp. U9/10 mossy heaths) & **snowbeds** (esp. U12), on **thin, stony soils** on **high ground** from Wales northwards (mainly in Scottish Highlands; also at lower altitudes in far N).

There is no single altitudinal threshold for the lower limit of a montane environment in Britain and Ireland. In southern areas these environments are restricted to the highest altitudes; for example, in Wales they are mostly >800 m. Further north they occur lower down. On Ronas Hill in Shetland montane heath extends down to well below 300 m in places, and a large area around the summit (450 m) has tundra-like habitat and vegetation similar to that found on the Cairngorms at >900 m.

Ronas Hill, Shetland

On relatively sheltered high slopes in the Scottish Highlands, especially where facing NW-E, snow lies well into summer in **snowbeds** whose vegetation (U11-14 & U18) consists mainly of mosses and liverworts. Few other plants can tolerate the prolonged snow lie but there is commonly a sparse scatter of snow-tolerant species such as Stiff Sedge, Dwarf Willow & Tufted Hair-grass.

A snowbed in the Ben Nevis range in midsummer; vegetation mainly mosses & liverworts

Kiaeria starkei - a snowbed moss

Snowbed vegetation: a mat of tiny grey-brown liverworts dotted with mosses, Stiff Sedge and Dwarf Willow

On ungrazed montane cliff ledges and other steep, ungrazed or lightly grazed montane slopes, are mixtures of grassland, heath, snowbeds (at high altitude) and tall herb communities.

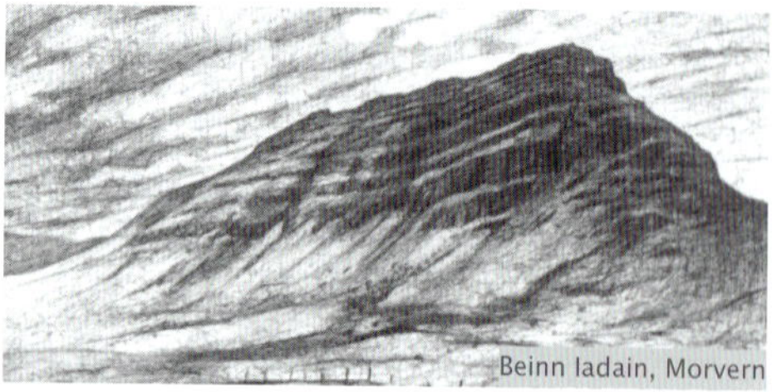

The **tall herb vegetation** includes lush assemblages of species that are palatable to sheep and deer and are much less common or luxuriant on more accessible grazed slopes. It varies from swards of Greater Woodrush (U16) on acid rocks and soils to mixtures of Greater Woodrush and herbs such as Angelica, Water Avens and Roseroot (U17) on basic ledges where the vegetation can be species-rich including some uncommon species.

Ungrazed tall herb vegetation on a montane cliff

Aonach Beag (Ben Nevis range)

The distinctive crinkly evergreen leaves (white underneath) of the rare **Mountain Avens** *Dryas octopetala* on basic cliffs; has white 8-petalled flowers (to 4 cm across) in summer.

Rocks

Rocks appear in the landscape in many places, from seashores to mountain tops. Rock outcrops occur where the underlying rock (**bedrock**) is exposed at the ground surface; where these are very steep they form **cliffs**. Large exposures of bedrock with intricate patterns of crevices (grikes) and intervening blocks (clints) are **pavements**: they are commonest in limestone areas in N England and W Ireland. Rock fragments broken off cliffs and rolled downslope, or that have been eroded and dumped by glaciers, now form accumulations of **boulders, stones and sand**. Below many cliffs are steep **screes** composed of unstable masses of angular rock fragments of varying sizes. Rounded, water-worn rock fragments carried along by rivers can be deposited as **river shingles**. Many types of vegetation - especially heaths and grasslands - occur patchily among rock habitats: on cliff ledges, in crevices, among boulders and screes, and on shingle. Rock surfaces support **small-scale assemblages of mosses, liverworts and lichens**. The vegetation of rocky habitats varies greatly, especially in relation to acidity, climate, shade and shelter. Much of it is of great value for conservation: for example, moss and liverwort communities in western woods, ravines and hillslopes.

Rocky landscape of Torridonian Sandstone in W Sutherland

Boulders of volcanic rock (Lake District)

Carboniferous Limestone in N Wales

Coastal habitats

Sea cliffs and rocky beaches form much of our coastline, especially in the north and west. Sea cliffs subjected to salt spray have specialized vegetation (MC1–12) with grasses such as Red Fescue and herbs such as Thrift and Sea Plantain, as well as other types of grassland, heath, scrub, woodland and bracken vegetation. Rocky beaches have algae (seaweeds) and coastal rocks can be good habitats for mosses, liverworts and lichens.

Near St Abbs, Scottish Borders: mostly grassland (NVC MC8/9 & U4)

Rocky coast with woodland, scrub and sheltered gullies; good for mosses, liverworts and lichens

Saltmarshes occupy muddy ground below the high tide line and so are mostly covered by seawater twice each day. Their vegetation (SM1–28) consists mainly of grasses (esp. Red Fescue, Common Saltmarsh-grass and Creeping Bent), rushes (esp. Saltmarsh Rush) and other species such as Sea Plantain, Sea Arrow-grass, Sea Aster, Sea Milkwort, Thrift and Glasswort, the last species mainly on barer muddy ground on the lower, seaward edges of the marsh or in depressions in the middle to upper marsh. Along the upper, less frequently inundated edges of the saltmarsh other species such as Couch, Tall Fescue, Silverweed, Autumn Hawkbit and Perennial Sow-thistle can be common.

Upper saltmarsh with Red Fescue, Creeping Bent, Common Saltmarsh-grass, Saltmarsh Rush, Thrift, Sea Plantain and Sea Milkwort

Strandlines and vegetated shingle at and just above the high tide line, at the seaward edges of sand dunes and the upper edges of saltmarshes and sandy, stony or rocky beaches, can have interesting vegetation (SD1–3) including oraches, Sea Mayweed, Sea Sandwort, Sea Rocket, Silverweed, Perennial Sow-thistle and Curled Dock.

Vegetated shingle/sand on the southern English coast

Sand dunes are areas of coastal sand, commonly forming patterns of low 'hills' and intervening hollows. On **'mobile' dunes** toward the seaward edge, much of the sand is loose and either bare or with pioneer colonizers such as Marram (SD6), Lyme-grass (SD5), Sand Couch (SD4) and Sand Sedge (SD10): species that spread by means of underground stems (rhizomes) and help to stabilize the sand. Further inland on more stable or **'fixed' dunes** (SD7–9 & SD12) Marram and Sand Sedge are mixed with other species including Red Fescue, False Oat-grass, Lady's Bedstraw, Common Bird's-foot-trefoil, Cat's-ear and Mouse-ear Hawkweed; some of these dunes in NW Scotland are made of calcareous shell sand and have herb-rich 'machair' grassland (SD8 & MG5).

Vegetation at the 'mobile' seaward edge of a dune complex in Sutherland; NVC SD6 and (lower R) SD4

Wet, periodically flooded hollows among dunes are **dune slacks**. Their wetland vegetation (SD13–17) has species including Common Sedge, Glaucous Sedge, Silverweed and Creeping Willow.

'Fixed' dune grassland and (in distance at right) heath. Dune slack in wetter (greener) hollow at right.

Habitats disturbed by human activity

In places such as roadsides, tracks, gateways, arable field margins and areas of urban and waste ground, artificial disturbance commonly prevents the vegetation from attaining complete cover of the ground. This disturbance benefits plant species that are not very good at competing with other plants for space: it prevents more competitive species from attaining dominance and swamping out poorer competitors. These poor competitors include species that many people call **weeds**: plants such as thistles, docks, Groundsel, Greater Plantain, Dandelion, Chickweed and dead-nettles. Most of common lowland weeds are species that grow best on dry to moist, fertile, more or less neutral soils. Some such as thistles, docks and Greater Plantain are tolerant of trampling; others such as dead-nettles and Groundsel are not tough enough to survive such treatment. Floristic variation in weedy communities (OV1-27) reflects differences in the type of soil and the nature of the disturbance. Weedy vegetation adds visual interest to our intensively-managed lowlands and provides habitats for insects, birds and mammals. Plants in these places commonly include non-native species, some of which are invasive.

In felled woodland on neutral to acid soils Rosebay Willowherb (OV27), Wavy Hair-grass (U2) or Foxglove can be abundant for a few years after felling. Rosebay Willowherb is also common in other disturbed places such as along railway lines. Wavy Hair-grass can be abundant in recently burned heath.

In areas of intensive agriculture, nutrients from manure and artificial fertilizers can accumulate at high concentrations in the soil. This eutrophication can encourage the growth of species such as Stinging Nettle (OV24-25), Creeping Buttercup and Chickweed.

One gradient of variation among artificially disturbed habitats is from places where disturbance is current or recent (vegetation typically short/sparse) to those where it took place longer ago (vegetation typically taller and thicker; these places can show fascinating aspects of wildness in the ways the vegetation has been able to develop with minimal intervention).

HABITAT CLASSIFICATIONS

UK National Vegetation Classification (NVC)

The most detailed British vegetation classification. Developed from analysis of sample plot data. Used for many purposes (e.g. surveys & conservation assessments). 286 communities; 681 categories if sub-communities are counted in. Descriptions, floristic tables, etc, in *British Plant Communities*, 5 vols.; edited by J.S. Rodwell (Cambridge University Press, 1991–2000). List of NVC community codes & names:

AQUATIC COMMUNITIES *(= Phase One F2 & G)*

A1 *Lemna gibba* community
A2 *Lemna minor* community
A3 *Spirodela polyrhiza – Hydrocharis morsus-ranae* comm.
A4 *Hydrocharis morsus-ranae – Stratiotes aloides* comm.
A5 *Ceratophyllum demersum* community
A6 *Ceratophyllum submersum* community
A7 *Nymphaea alba* community
A8 *Nuphar lutea* community
A9 *Potamogeton natans* community
A10 *Persicaria amphibia* community
A11 *Stuckenia pectinata – Myriophyllum spicatum* community
A12 *Stuckenia pectinata* community
A13 *Potamogeton perfoliatus – Myriophyllum alterniflorum* community
A14 *Myriophyllum alterniflorum* community
A15 *Elodea canadensis* community
A16 *Callitriche stagnalis* community
A17 *Ranunculus penicillatus* ssp. *pseudofluitans* comm.
A18 *Ranunculus fluitans* community
A19 *Ranunculus aquatilis* community
A20 *Ranunculus peltatus* community
A21 *Ranunculus baudotii* community
A22 *Littorella uniflora – Lobelia dortmanna* community
A23 *Isoetes lacustris/echinospora* community
A24 *Juncus bulbosus* community

CALCAREOUS GRASSLANDS *(= Phase One B3)*

CG1 *Festuca ovina – Carlina vulgaris* grassland
CG2 *Festuca ovina – Helictochloa pratensis* grassland
CG3 *Bromopsis erecta* grassland
CG4 *Brachypodium pinnatum* grassland
CG5 *Bromopsis erecta – Brachypodium pinnatum* grassland
CG6 *Helictotrichon pubescens* grassland
CG7 *Festuca ovina – Pilosella officinarum – Thymus drucei/pulegioides* grassland
CG8 *Sesleria caerulea – Scabiosa columbaria* grassland
CG9 *Sesleria caerulea – Galium sterneri* grassland
CG10 *Festuca ovina – Agrostis capillaris – Thymus drucei* grassland
CG11 *F. ovina–A. capillaris–Alchemilla alpina* grass-heath
CG12 *F. ovina–A. alpina–Silene acaulis* dwarf-herb comm.
CG13 *Dryas octopetala – Carex flacca* heath
CG14 *Dryas octopetala – Silene acaulis* ledge community

HEATHS *(= Phase One D)*

H1 *Calluna vulgaris – Festuca ovina* heath
H2 *Calluna vulgaris – Ulex minor* heath
H3 *Ulex minor – Agrostis curtisii* heath
H4 *Ulex gallii – Agrostis curtisii* heath
H5 *Erica vagans – Schoenus nigricans* heath
H6 *Erica vagans – Ulex europaeus* heath
H7 *Calluna vulgaris – Scilla verna* heath
H8 *Calluna vulgaris – Ulex gallii* heath
H9 *Calluna vulgaris – Avenella flexuosa* heath
H10 *Calluna vulgaris – Erica cinerea* heath
H11 *Calluna vulgaris – Carex arenaria* heath
H12 *Calluna vulgaris – Vaccinium myrtillus* heath
H13 *Calluna vulgaris – Cladonia arbuscula* heath
H14 *Calluna vulgaris – Racomitrium lanuginosum* heath
H15 *Calluna vulgaris – Juniperus communis* ssp. *nana* heath
H16 *Calluna vulgaris – Arctostaphylos uva-ursi* heath
H17 *Calluna vulgaris – Arctostaphylos alpinus* heath
H18 *Vaccinium myrtillus – Avenella flexuosa* heath
H19 *Vaccinium myrtillus – Cladonia arbuscula* heath
H20 *Vaccinium myrtillus – Racomitrium lanuginosum* heath
H21 *Calluna – V. myrtillus – Sphagnum capillifolium* heath
H22 *Vaccinium myrtillus – Rubus chamaemorus* heath

MIRES *(= Phase One B5, D2 & E)*

M1 *Sphagnum auriculatum* bog pool community
M2 *Sphagnum cuspidatum/fallax* bog pool community
M3 *Eriophorum angustifolium* bog pool community
M4 *Carex rostrata – Sphagnum fallax* mire
M5 *Carex rostrata – Sphagnum squarrosum* mire
M6 *Carex echinata – Sphagnum fallax/auriculatum* mire
M7 *Carex canescens – Sphagnum russowii* mire
M8 *Carex rostrata – Sphagnum warnstorfii* mire
M9 *Carex rostrata – Calliergonella cuspidata/Calliergon giganteum* mire
M10 *Carex dioica – Pinguicula vulgaris* mire
M11 *Carex demissa – Saxifraga aizoides* mire
M12 *Carex saxatilis* mire
M13 *Schoenus nigricans – Juncus subnodulosus* mire
M14 *Schoenus nigricans – Narthecium ossifragum* mire
M15 *Trichophorum germanicum – Erica tetralix* wet heath
M16 *Erica tetralix – Sphagnum compactum* wet heath
M17 *Trichophorum germanicum – Eriophorum vaginatum* blanket mire
M18 *Erica tetralix – Sphagnum papillosum* raised & blanket mire
M19 *Calluna vulgaris – Eriophorum vaginatum* blanket mire
M20 *Eriophorum vaginatum* raised and blanket mire
M21 *Narthecium ossifragum – Sphagnum papillosum* mire
M22 *Juncus subnodulosus – Cirsium palustre* fen-meadow
M23 *Juncus effusus/acutiflorus – Galium palustre* rush-pasture
M24 *Molinia caerulea – Cirsium dissectum* fen-meadow
M25 *Molinia caerulea – Potentilla erecta* mire
M26 *Molinia caerulea – Crepis paludosa* mire
M27 *Filipendula ulmaria – Angelica sylvestris* mire
M28 *Iris pseudacorus – Filipendula ulmaria* mire
M29 *Hypericum elodes – Potamogeton polygonifolius* soakway
M30 Related vegetation of seasonally-inundated habitats
M31 *Anthelia julacea – Sphagnum auriculatum* spring
M32 *Philonotis fontana – Saxifraga stellaris* spring
M33 *Pohlia wahlenbergii* var. *glacialis* spring
M34 *Carex demissa – Koenigia islandica* flush
M35 *Ranunculus omiophyllus – Montia fontana* rill
M36 Lowland springs & streambanks of shaded situations
M37 *Palustriella commutata – Festuca rubra* spring
M38 *Palustriella commutata – Carex nigra* spring

MARITIME CLIFF COMMUNITIES *(= Phase One H4, H5, H8)*

MC1 *Crithmum maritimum - Spergularia rupicola* maritime rock-crevice community
MC2 *Armeria maritima - Ligusticum scoticum* maritime rock-crevice community
MC3 *Rhodiola rosea - Armeria maritima* maritime cliff-ledge community
MC4 *Brassica oleracea* maritime cliff-ledge community
MC5 *Armeria maritima - Cerastium diffusum* maritime therophyte community
MC6 *Atriplex prostrata - Beta vulgaris* ssp. *maritima* sea-bird cliff community
MC7 *Stellaria media - Rumex acetosa* sea-bird cliff comm.
MC8 *Festuca rubra - Armeria maritima* maritime grassland
MC9 *Festuca rubra - Holcus lanatus* maritime grassland
MC10 *Festuca rubra - Plantago* spp. maritime grassland
MC11 *Festuca rubra - Daucus carota* ssp. *gummifer* maritime grassland
MC12 *Festuca rubra - Hyacinthoides non-scripta* maritime bluebell community

MESOTROPHIC GRASSLANDS *(= Phase One B2/B4-6)*

MG1 *Arrhenatherum elatius* grassland
MG2 *Arrhenatherum elatius - Filipendula ulmaria* grassland
MG3 *Anthoxanthum odoratum - Geranium sylvaticum* grassland
MG4 *Alopecurus pratensis - Sanguisorba officinalis* grassland
MG5 *Cynosurus cristatus - Centaurea nigra* grassland
MG6 *Lolium perenne - Cynosurus cristatus* grassland
MG7 *Lolium perenne* leys and related grasslands
MG8 *Cynosurus cristatus - Caltha palustris* grassland
MG9 *Holcus lanatus - Deschampsia cespitosa* grassland
MG10 *Holcus lanatus - Juncus effusus* rush-pasture
MG11 *Festuca rubra - Agrostis stolonifera - Potentilla anserina* grassland
MG12 *Schedonorus arundinaceus* grassland
MG13 *Agrostis stolonifera - Alopecurus geniculatus* grassland
MG14 *Carex nigra-Agrostis stolonifera-Jacobaea aquatica* grassland
MG15 *Alopecurus pratensis-Poa trivialis-Cardamine pratensis* grassland
MG16 Provisional *Agrostis stolonifera-Eleocharis palustris* inundation grassland

SWAMPS AND FENS *(= Phase One F1 & F2)*

S1 *Carex elata* sedge-swamp
S2 *Cladium mariscus* swamp and sedge-beds
S3 *Carex paniculata* swamp
S4 *Phragmites australis* swamp and reed-beds
S5 *Glyceria maxima* swamp
S6 *Carex riparia* swamp
S7 *Carex acutiformis* swamp
S8 *Schoenoplectus lacustris* swamp
S9 *Carex rostrata* swamp
S10 *Equisetum fluviatile* swamp
S11 *Carex vesicaria* swamp
S12 *Typha latifolia* swamp
S13 *Typha angustifolia* swamp
S14 *Sparganium erectum* swamp
S15 *Acorus calamus* swamp
S16 *Sagittaria sagittifolia* swamp
S17 *Carex pseudocyperus* swamp
S18 *Carex otrubae* swamp
S19 *Eleocharis palustris* swamp
S20 *Schoenoplectus tabernaemontani* swamp
S21 *Bolboschoenus maritimus* swamp
S22 *Glyceria fluitans* water-margin vegetation
S23 Other water-margin vegetation
S24 *Phragmites australis - Thysselinum palustre* fen
S25 *Phragmites australis - Eupatorium cannabinum* fen
S26 *Phragmites australis - Urtica dioica* tall-herb fen
S27 *Carex rostrata - Comarum palustre* tall-herb fen
S28 *Phalaris arundinacea* tall-herb fen

SAND DUNE COMMUNITIES *(= Phase One H5 & H6)*

SD1 *Rumex crispus - Glaucium flavum* shingle community
SD2 *Honckenya peploides - Cakile maritima* strandline community
SD3 *Tripleurospermum maritimum - Galium aparine* strandline community
SD4 *Elymus junceiformis* foredune community
SD5 *Leymus arenarius* mobile dune community
SD6 *Ammophila arenaria* mobile dune community
SD7 *Ammophila arenaria - Festuca rubra* semi-fixed dune
SD8 *Festuca rubra - Galium verum* fixed dune grassland
SD9 *Ammophila arenaria - Arrhenatherum elatius* dune grassland
SD10 *Carex arenaria* dune community
SD11 *Carex arenaria - Cetraria aculeata* dune comm.
SD12 *Carex arenaria - Festuca ovina - Agrostis capillaris* dune grassland
SD13 *Sagina nodosa - Bryum pseudotriquetrum* dune-slack
SD14 *Salix repens - Campylium stellatum* dune-slack
SD15 *Salix repens - Calliergonella cuspidata* dune-slack
SD16 *Salix repens - Holcus lanatus* dune-slack community
SD17 *Potentilla anserina - Carex nigra* dune-slack
SD18 *Hippophae rhamnoides* dune scrub
SD19 *Phleum arenarium - Arenaria serpyllifolia* dune annual community

SALTMARSH COMMUNITIES *(= Phase One H1 & H2)*

SM1 *Zostera* communities
SM2 *Ruppia maritima* salt-marsh community
SM3 *Eleocharis parvula* salt-marsh community
SM4 *Spartina maritima* salt-marsh community
SM5 *Spartina alterniflora* salt-marsh community
SM6 *Spartina anglica* salt-marsh community
SM7 *Sarcocornia perennis* stands
SM8 Annual *Salicornia* salt-marsh community
SM9 *Suaeda maritima* salt-marsh community
SM10 Transitional low-marsh vegetation with *Puccinellia maritima*, annual *Salicornia* spp. & *Suaeda maritima*
SM11 *Tripolium pannonicum* var. *discoideus* salt-marsh comm.
SM12 Rayed *Tripolium pannonicum* on salt-marshes
SM13 *Puccinellia maritima* salt-marsh community
SM14 *Atriplex portulacoides* salt-marsh community
SM15 *Juncus maritimus - Triglochin maritima* salt-marsh
SM16 *Festuca rubra* salt-marsh community
SM17 *Artemisia maritima* salt-marsh community
SM18 *Juncus maritimus* salt-marsh community
SM19 *Blysmus rufus* salt-marsh community
SM20 *Eleocharis uniglumis* salt-marsh community
SM21 *Suaeda vera - Limonium binervosum* salt-marsh
SM22 *Atriplex portulacoides - Frankenia laevis* saltmarsh
SM23 *Spergularia marina - Puccinellia distans* saltmarsh
SM24 *Elymus athericus* salt-marsh community
SM25 *Suaeda vera* drift-line community
SM26 *Limbarda crithmoides* on salt-marshes
SM27 Ephemeral salt-marsh veg. with *Sagina maritima*
SM28 *Elymus repens* salt-marsh community

OPEN VEGETATION COMMUNITIES *(= Phase One A4,C3,F2,I,J)*

OV1 *Viola arvensis – Aphanes australis* community
OV2 *Briza minor – Silene gallica* community
OV3 *Papaver rhoeas – Viola arvensis* community
OV4 *Glebionis segetum – Spergula arvensis* comm.
OV5 *Digitaria ischaemum – Erodium cicutarium* community
OV6 *Cerastium glomeratum – Fumaria muralis* ssp. *boraei* community
OV7 *Veronica persica – Veronica polita* community
OV8 *Veronica persica – Alopecurus myosuroides* comm.
OV9 *Tripleurospermum inodorum– Stellaria media* comm.
OV10 *Poa annua – Senecio vulgaris* community
OV11 *Poa annua – Stachys arvensis* community
OV12 *Poa annua – Myosotis arvensis* community
OV13 *Stellaria media – Capsella bursa-pastoris* community
OV14 *Urtica urens – Lamium amplexicaule* community
OV15 *Lysimachia arvensis – Veronica persica* community
OV16 *Papaver rhoeas – Silene noctiflora* community
OV17 *Reseda lutea – Polygonum aviculare* community
OV18 *Polygonum aviculare – Matricaria discoidea* comm.
OV19 *Poa annua – Tripleurospermum inodorum* comm.
OV20 *Poa annua – Sagina procumbens* community
OV21 *Poa annua – Plantago major* community
OV22 *Poa annua – Taraxacum officinale* community
OV23 *Lolium perenne – Dactylis glomerata* community
OV24 *Urtica dioica – Galium aparine* community
OV25 *Urtica dioica – Cirsium arvense* community
OV26 *Epilobium hirsutum* community
OV27 *Chamaenerion angustifolium* community
OV28 *Agrostis stolonifera – Ranunculus repens* community
OV29 *Alopecurus geniculatus – Rorippa palustris* comm.
OV30 *Bidens tripartita – Persicaria amphibia* community
OV31 *Rorippa palustris – Gnaphalium uliginosum* comm.
OV32 *Myosotis scorpioides – Ranunculus sceleratus* comm.
OV33 *Persicaria lapathifolia – Poa annua* community
OV34 *Allium schoenoprasum – Plantago maritima* comm.
OV35 *Lythrum portula – Ranunculus flammula* community
OV36 *Lythrum hyssopifolia – Juncus bufonius* community
OV37 *Festuca ovina – Minuartia verna* community
OV38 *Gymnocarpium robertianum – Arrhenatherum elatius* community
OV39 *Asplenium trichomanes – Asplenium ruta-muraria* community
OV40 *Asplenium viride – Cystopteris fragilis* community
OV41 *Parietaria judaica* community
OV42 *Cymbalaria muralis* community

ACID GRASSLAND & UPLAND COMMUNITIES *(P.OneB1,C,D3-6)*

U1 *Festuca ovina – Agrostis capillaris – Rumex acetosella* grassland
U2 *Avenella flexuosa* grassland
U3 *Agrostis curtisii* grassland
U4 *Festuca ovina – Agrostis capillaris – Galium saxatile* grassland
U5 *Nardus stricta – Galium saxatile* grassland
U6 *Juncus squarrosus – Festuca ovina* grassland
U7 *Nardus stricta – Carex bigelowii* grass-heath
U8 *Carex bigelowii – Polytrichastrum alpinum* sedge-heath
U9 *Juncus trifidus – Racomitrium lanuginosum* rush-heath
U10 *Carex bigelowii – Racomitrium lanuginosum* moss-heath
U11 *Polytrichum sexangulare – Kiaeria starkei* snow-bed
U12 *Salix herbacea – Racomitrium heterostichum* snow-bed
U13 *Deschampsia cespitosa – Galium saxatile* grassland
U14 *Alchemilla alpina – Sibbaldia procumbens* dwarf-herb community
U15 *Saxifraga aizoides – Alchemilla glabra* banks
U16 *Luzula sylvatica – Vaccinium myrtillus* tall-herb community
U17 *Luzula sylvatica – Geum rivale* tall-herb community
U18 *Cryptogramma crispa – Athyrium distentifolium* snow-bed
U19 *Oreopteris limbosperma–Blechnum spicant* community
U20 *Pteridium aquilinum – Galium saxatile* community
U21 *Cryptogramma crispa – Avenella flexuosa* community

WOODLAND & SCRUB COMMUNITIES *(= Phase One A, C1/3)*

W1 *Salix cinerea – Galium palustre* woodland
W2 *Salix cinerea – Betula pubescens – Phragmites australis* woodland
W3 *Salix pentandra – Carex rostrata* woodland
W4 *Betula pubescens – Molinia caerulea* woodland
W5 *Alnus glutinosa – Carex paniculata* woodland
W6 *Alnus glutinosa – Urtica dioica* woodland
W7 *Alnus – Fraxinus – Lysimachia nemorum* woodland
W8 *Fraxinus excelsior – Acer campestre – Mercurialis perennis* woodland
W9 *Fraxinus excelsior – Sorbus aucuparia – Mercurialis perennis* woodland
W10 *Quercus robur – Pteridium aquilinum – Rubus fruticosus* woodland
W11 *Q. petraea – B. pubescens – Oxalis acetosella* woodl.
W12 *Fagus sylvatica – Mercurialis perennis* woodland
W13 *Taxus baccata* woodland
W14 *Fagus sylvatica – Rubus fruticosus* woodland
W15 *Fagus sylvatica – Avenella flexuosa* woodland
W16 *Quercus –Betula – Avenella flexuosa* woodland
W17 *Q. petraea – B. pubescens – Dicranum majus* woodland
W18 *Pinus sylvestris – Hylocomium splendens* woodland
W19 *Juniperus communis – Oxalis acetosella* woodland
W20 *Salix lapponum – Luzula sylvatica* scrub
W21 *Crataegus monogyna – Hedera helix* scrub
W22 *Prunus spinosa – Rubus fruticosus* scrub
W23 *Ulex europaeus – Rubus fruticosus* scrub
W24 *Rubus fruticosus – Holcus lanatus* underscrub
W25 *Pteridium aquilinum – Rubus fruticosus* underscrub

Additional vegetation types that cannot clearly be fitted into the current NVC include the following:

Most conifer plantations (esp. dense plantations with species-poor ground flora – v. common in N & W)
Oak/birchwood + virtually pure *Luzula sylvatica* swards
Pinus–Cladonia lichen-rich pinewood (NE Scotland)
Elder scrub (related to W21)
Raspberry underscrub (related to W24)
Neutral sedge flushes with *Carex panicea/nigra/echinata* and various herbs & mosses (like 'non-rushy M23')
Equisetum palustre-dominated mire (like 'horsetail M23')
Swamps/fens of *Carex acuta, C. aquatilis* or *C. lasiocarpa*
Species-poor prostrate *Calluna* heath (mainly montane)
Non-montane *Vaccinium* heath + *Sphagnum capillifolium*
Dryopteris affinis agg.-dominated vegetation
Juncus effusus vegetation + acid grassland spp. (*Potentilla erecta, Galium saxatile* etc)
Juncus acutiflorus swards + acid grassland spp.
Carex nigra swards + acid grassland spp.
Species-poor grasslands dominated by *Holcus lanatus*, by *H. mollis* or by *H. lanatus/Festuca rubra*
Petasites hybridus (Butterbur) – dominated vegetation
Impatiens glandulifera (Himalayan Balsam)-dominated veg.
Reynoutria japonica (Japanese Knotweed)-dominated veg.
Cereal and vegetable crops

Phase One habitat classification

A broad habitat classification developed and used in Great Britain since the 1970s. The main structure is shown below (with NVC equivalents). Further subdivisions are omitted here because of limited space. The Phase One manual is available at www.jncc.gov.uk.

Code	Phase One name	**Main NVC equivalents** (note that some NVC communities can belong within two or more Phase One habitats)
A1.1	Broadleaved woodland	W2-12, W14-17
A1.2	Coniferous woodland	W13, W18, some plantation forms of W10-11, W16 & W18, & 'non-NVC' conifer plantation
A1.3	Mixed woodland	Some forms of W8-18
A2	Scrub	W19-23, W1 & forms of W2-7
A3	Parkland/scattered trees	Includes a varied range of NVC types (e.g. MG6/7)
A4	Recently felled woodland	Can include W24, U2 & OV27
B1	Acid grassland	U1-6 & inland forms of SD10-11
B2	Neutral grassland	MG1-5, MG8-10, MG12-13
B3	Calcareous grassland	CG1-14, OV37
B4	Improved grassland	MG6-7
B5	Marsh/marshy grassland	MG8, MG10, M22-28
B6	Poor semi-improved grassland	Some examples of MG6 and of other grassland communities
C1	Bracken - continuous	U20, W25
C2	Upland species-rich ledges	U17
C3.1	Other tall herb & fern: ruderal	OV24-27, W24
C3.2	Other tall herb/fern: non-ruderal	U16, U18-19, U21, OV38
D1	Dry dwarf shrub heath	H1-4, H6-10, H12-22
D2	Wet dwarf shrub heath	M15-16, H5
D3	Lichen/bryophyte heath	Some H1, H13-14, H17, H19-20, U1, U10, SD11
D4	Montane heath/dwarf herb	U7-15
D5	Dry heath/acid grassland mosaic	Mix of B1 and D1 communities
D6	Wet heath/acid grassland mosaic	Mix of B1 and D2 communities
E1	Bog	M1-3, M17-20 (& M15-16 & M25 on deep peat)
E2	Flush & spring	M4, M6-14, M21, M29, M31-38
E3	Fen	M1-9, M21, M29, S24-28 and other mires & swamps
E4	Bare peat	
F1	Swamp	S1-22, S27-28
F2	Water margin & inundation veg.	S1-23/26-28, A1-3/10/16/19-21, M30, MG11/13, OV28-36
G1	Standing water	Incl. A1-17, A19-20
G2	Running water	Incl. A1-2, A5-9, A11-21, SM2
H1	Intertidal	Includes SM1-3
H2	Saltmarsh	SM4-28, MG11-12
H3	Shingle above high tide mark	SD1-4, MC6 & some maritime grassland, heath & scrub
H4	Rocks above high tide mark	MC1-7
H5	Strandline vegetation	SD1-3, MC6
H6	Sand dune	SD4-19 and some U1, H1, H10-11, M15-16
H8	Maritime cliff and slope	MC1-12 and some H7-8, SM16
I1	Natural rock exposure and waste	Includes U18, U21, OV38-41
I2	Artificial rock exposure & waste	Includes OV39
J1	Cultivated/disturbed land	Includes OV1-23
J2	Boundaries (hedges, fences etc)	Includes OV39/42
J3	Built-up areas	Includes OV20-25/27
J4	Bare ground	
J5	Other habitats	

UK Biodiversity Action Plan (UK BAP) habitats

All UK habitats can be assigned to one of the **Broad Habitats** (1–21 below), most of which include one or more **Priority Habitats** of ecological importance (in italics below; main NVC communities of terrestrial Priority Habitats in brackets; some NVC types in two or more Priority Habitats.) There are also Priority Species (e.g. rarities). See www.jncc.gov.uk. In Scotland, Priority habitats/species are **Scottish Biodiversity List (SBL)** habitats/species; SBL now revised; c.80 habitats (based on EUNIS – see p. 383); https://www.gov.scot/publications/scottish-biodiversity-list/

1 **Broadleaved, mixed & yew woodland:** All woodland and scrub (including plantation) that is not coniferous. *Includes eight Priority Habitats: Lowland mixed deciduous woodland (W8/10/16), Lowland beech and yew woodland (W12–15), Upland birchwoods (W11/17 & some W4a), Upland mixed ashwoods (W7c/W9 & some W8), Upland oakwood (W11/17 & some W4a), Wet woodland (W1–7), Wood pasture & parkland (mature open-grown trees scattered among vegetation of various NVC types such as grassland that is or has been grazed) & Traditional orchards (that have larger, more well-spaced trees than commercial orchards and are not sprayed with herbicide; includes various NVC types).*

2 **Coniferous woodland:** All coniferous woodland except that of yew. Includes all conifer plantations. *Includes one Priority Habitat: Native pinewoods (W18 pinewood & W19 juniper scrub).*

3 **Boundary and linear features:** Narrow, linear features such as hedges and walls. Also non-urban roads, tracks and railways. *Includes one Priority Habitat: Ancient and/or species rich hedgerows (includes some MG1 and W21–24).*

4 **Arable and horticulture:** Arable land, commercial orchards, nurseries and other commercial horticultural land, ploughed fields and set-aside land. *Includes one Priority Habitat: Cereal field margins (incl. MG1, OV1/3/4/7/9–10/12–14, W24).*

5 **Improved grassland:** Agriculturally treated, intensively managed grasslands (NVC MG6–7); see description of Improved grassland on page 355. *This Broad Habitat does not include any Priority Habitats.*

6 **Neutral grassland:** Less intensively managed grasslands on neutral soils (NVC MG1–5 & 8–13); see description on page 353. *Includes the following Priority Habitats: Lowland meadows (MG4/5/8) & Upland hay meadow (MG3).*

7 **Calcareous grassland:** See description on page 354. NVC: CG1–14. *The Priority Habitats Lowland calcareous grassland (CG1–8 & some CG9–10) and Upland calcareous grassland (CG9–14) make up the whole of this Broad Habitat.*

8 **Acid grassland:** See description of Acid grassland on page 352. NVC: U1–6 and inland SD10–12. *Includes one Priority Habitat: Lowland dry acid grassland (U1–4 in lowland situations, and inland SD10–12).*

9 **Bracken:** Vegetation dominated by Bracken (NVC U20 & W25) (see page 351). *No Priority Habitats except on sea cliffs (see below under Supralittoral Rock).*

10 **Dwarf shrub heath:** All heaths (see page 356) except montane ones (see under 15 Montane habitats) and wet heaths (M15/16) on peat >50 cm deep (see under 12 Bog). *The Priority Habitats Lowland heathland (within agricultural lowlands – mainly H1–9 & some H10/12/21/M15/16) & Upland heathland (in unenclosed upland areas – mainly H10/12/16/18/21 & M15/16, & some H4/8/9) make up the whole of this Broad Habitat.*

11 **Fen, marsh & swamp:** All wetlands (see page 360) except bogs. *Includes five Priority Habitats: Lowland fens (occurrences within agricultural lowlands, of M1–6/9–10/13–14/21/27–30/36, S1–3/5–23/28 & examples of S24–27 not dominated by Phragmites), Upland flushes, fens & swamps (in unenclosed upland areas (occurrences in unenclosed upland areas, of M4–12/21/23a/25c/27–29/31–35/37–38, S3/9–11/19 and examples of S27 not dominated by Phragmites), Reedbeds (Phragmites-dominated S4 & S24–27), Purple moor-grass & rush pasture (Most M26 and most lowland examples of M22–25), and Coastal & floodplain grazing marsh (M22–25, MG4/6–13 and S5–7 in grazed, periodically inundated coastal marshes).*

12 **Bog:** All Bogs (NVC 1–3 & 17–21) – see page 360 – and also examples of wet heath (M15–16; page 356) and vegetation dominated by Purple moor-grass (M25; page 366) that are on wet peat more than 50 cm deep. *Includes two Priority Habitats: Blanket bog (mainly in upland areas of N & W) and Lowland raised bog (in raised domes of peat in lowland areas); raised peat domes within larger areas of blanket bog are generally included in 'Blanket bog'.*

13 **Standing open water and canals:** Lakes, lochs, ponds, reservoirs, canals, water-filled ditches and flooded gravel pits, but not small bog pools (= Bog broad habitat) and temporarily flooded areas. *Includes several aquatic Priority Habitats.*

14 **Rivers and streams:** Rivers and streams, including adjacent steep banks or margins within the mean annual flood zone. *Includes several aquatic Priority Habitats.*

15 **Montane habitats:** High altitude habitats (see page 370) whose vegetation (NVC U7–15, W20, H13–15/17/19/20/22 and high-altitude H12/18/21, M15 & U4–6) includes montane plant species such as Stiff sedge, Dwarf willow, the montane form of Crowberry (*Empetrum nigrum* ssp. *hermaphroditum*) or prostrate, wind-pruned Heather. *The Priority Habitat Mountain heaths and willow scrub makes up the whole of this Broad Habitat.*

16 **Inland rock:** Inland cliffs, screes, quarries, limestone pavements, excavated ground (e.g. opencast coal mines) and waste tips (such as those associated with mines). *Includes the following Priority Habitats: Inland Rock Outcrop and Scree Habitats, Limestone pavements, Calaminarian grasslands and Open mosaic habitats on previously developed land.*

17 **Built-up areas & gardens:** Towns, villages (including parks), farm steadings, allotments, gardens, industrial estates, shopping centres, airports, waste ground and other man-made structures, but not amenity grassland (= Improved grassland) or roads, tracks & railways in rural locations (= Boundary & linear features). *No Priority Habitats.*

18 **Supralittoral rock:** Coastal cliffs & steep slopes (see page 374). *Includes the Priority Habitat Maritime cliff and slope (MC1–12 & sea cliff examples of other types of grassland, heath, bracken & scrub).*

19 **Supralittoral sediment:** Coastal sand dunes and shingly strandline habitats (see pages 375–376). *Includes the Priority Habitats Coastal sand dunes, Coastal vegetated shingle and Machair.*

20 **Littoral rock:** Rocky beaches covered at high tide. *Includes several marine Priority Habitats.*

21 **Littoral sediment:** Sandy or muddy areas covered at high tide; includes saltmarshes (see page 375). *Includes the Priority Habitat Coastal Saltmarsh and several marine Priority Habitats.*

Annex I habitats of the European Union Habitats Directive

These are of special interest on a European scale, for which reason examples at many designated sites (e.g. Special Areas of Conservation) are given special protection. Asterisks indicate 'priority' status: this is not related to UK BAP/SBL Priority Habitats (page 382). This list also gives NVC communities occurring in terrestrial Annex 1 types; some NVC types occur in two or more Annex 1 types. More information at www.jncc.gov.uk

Marine, coastal and halophytic habitats

H1110 Sandbanks slightly covered by sea water
H1130 Estuaries
H1140 Mudflats & sandflats not covered at low tide
H1150 Coastal lagoons *
H1160 Large shallow inlets and bays
H1170 Reefs
H1210 Annual vegetation of drift lines *(MC6, SD2-3)*
H1220 Perennial vegetation of stony banks *(incl. SD1)*
H1230 Vegetated Atlantic/Baltic sea cliffs *(H7-8, MC1-12)*
H1310 Salicornia & other annuals on mud/sand *(SM7-9/27)*
H1320 Spartina swards (*Spartinion maritimae*) *(SM4-6)*
H1330 Atlantic salt meadows *(SM10-20)*
H1340 Inland salt meadows * *(inland SM16/23)*
H1420 Medit. & thermo-Atl. halophilous scrub *(SM7/21/25)*

Coastal sand dunes and continental dunes

H2110 Embryonic shifting dunes *(SD2/4/5)*
H2120 Shifting dunes with *Ammophila arenaria (SD5-6)*
H2130 Fixed dunes with herbaceous vegetation * *(SD7-9/11-12/19, CG10)*
H2140 Decalcified fixed dunes with *Empetrum* * *(H11)*
H2150 Atlantic decalcified fixed dunes * *(H1, H11)*
H2160 Dunes with *Hippophae rhamnoides (SD18)*
H2170 Dunes with *Salix repens* ssp. *argentea (SD16)*
H2190 Humid dune slacks *(SD13-17)*
H21A0 Machairs *(SD7-8, SD17, MG11, SD19)*
H2250 Coastal dunes with *Juniperus* spp. *
H2330 Inland dunes with *Corynephorus/Agrostis (SD11-12)*

Freshwater habitats

H3110 Mineral-poor oligotrophic waters of sandy plains
H3130 Oligotrophic to mesotrophic standing waters
H3140 Hard oligo-mesotrophic waters with *Chara* spp.
H3150 Natural eutrophic lakes with *Magnopotamion* or *Hydrocharition*-type vegetation
H3160 Natural dystrophic lakes and ponds
H3170 Mediterranean temporary ponds *
H3180 Turloughs *
H3260 Watercourses with *Ranunculion fluitantis* and *Callitricho-Batrachion* vegetation

Temperate heath and scrub

H4010 N Atlantic wet heaths with *Erica tetralix (H5, M14-16)*
H4020 Temperate Atlantic wet heaths with *Erica ciliaris* and *Erica tetralix* * *(H3-4, M16/21)*
H4030 European dry heaths *(H1-4/7-10/12/16/18/21)*
H4040 Dry Atlantic coastal heaths with *Erica vagans* * *(H6)*
H4060 Alpine and Boreal heaths *(H10/12-22)*
H4080 Sub-Arctic *Salix* spp. scrub *(W20)*

Sclerophyllous scrub (matorral)

H5110 Stable *Buxus sempervirens* on rocky slopes
H5130 *Juniperus* on heaths/calcareous grasslands *(incl. W19)*

Natural and semi-natural grassland formations

H6130 Calaminarian grasslands *(OV37)*
H6150 Siliceous alpine and boreal grasslands *(U7-12, U14)*
H6170 Alpine and subalpine calcareous grasslands *(CG12-14)*
H6210 Semi-natural dry grassland on calcareous substrates (incl. H6211: important orchid sites*) *(CG1-10)*
H6230 Species-rich *Nardus* grassland on siliceous substrates in mountain areas * *(CG10-11 & some U4-5)*
H6410 Molinia meadows on calcareous, peaty or clayey-silt-laden soils (*Molinion caeruleae*) *(M24/26)*
H6430 Hydrophilous tall herb fringe communities of plains and of the montane to alpine levels *(U17)*
H6510 Lowland hay meadows (*Alopecurus pratensis, Sanguisorba officinalis*) *(MG4)*
H6520 Mountain hay meadows *(MG3)*

Raised bogs and mires and fens

H7110 Active raised bogs * *(M1, M2, M18, M19)*
H7120 Degraded raised bogs still capable of natural regeneration *(M3, M15, M16, M18, M20, M25)*
H7130 Blanket bogs * *(M1-3, M17-20 & deep peat M15/25)*
H7140 Transition mire/quaking bog *(M4/5/8/9/14/29, S27)*
H7150 Depressions on peat substrates of the *Rhynchosporion (M1/2/17-19/21)*
H7210 Calcareous fens with *Cladium* & other sedges * *(M9/13-14/24, S2/24-25, SD14-15)*
H7220 Petrifying springs with tufa formation * *(M37, M38)*
H7230 Alkaline fens *(M9, M10, M13)*
H7240 Alpine *Caricion bicoloris-atrofuscae* * *(M10-12)*

Rocky habitats and caves

H8110 Siliceous scree of the montane to snow levels *(U18/21)*
H8120 Calcareous/calcshist montane/alpine scree *(OV38/40)*
H8210 Calcareous rocky slopes with chasmophytic vegetation *(OV39-40)*
H8220 Siliceous rocky slopes with chasmophytic veg. *(U21)*
H8240 Limestone pavement * *(CG9-10,OV38-40,W8-9/13/19/21)*
H8310 Caves not open to the public
H8330 Submerged or partially submerged sea caves

Forests

H9120 Atlantic acidophilous beech forests *(W14/15)*
H9130 Asperulo-Fagetum beech forests *(W12/14)*
H9160 Sub-Atl./medio-European oak/hornbeam forest *(W10)*
H9180 *Tilio-Acerion* forest of slopes/screes/ravines * *(W8-9)*
H9190 Old acidophilous oak woods on sandy plains *(W10/16)*
H91A0 Old sessile oakwoods *(W10/11/16/17)*
H91C0 Caledonian forest * *(W18 & some W4 & W17)*
H91D0 Bog woodland * *(some W2/3/4/18)*
H91E0 Alluvial forests with *Alnus* & *Fraxinus* * *(W2/5/6/7)*
H91J0 *Taxus baccata* woods of the British Isles *(W13)*

European Nature Information System (EUNIS)

A classification of all habitat types in Europe. Forms the basis of Scottish Biodiversity List Priority Habitats (see page 382). See https://www.eea.europa.eu/data-and-maps/data/eunis-habitat-classification-1/folder_contents and NatureScot Commissioned Report 766: Manual of terrestrial EUNIS habitats in Scotland; ISBN 978-1-78391-164-6, available at https://www.nature.scot/doc/naturescot-commissioned-report-766-manual-terrestrial-eunis-habitats-scotland (incl. EUNIS/NVC conversion).

UK Habitat Classification (UKHab)

A recent classification, available since 2018, covering all British habitats. Structured with five levels: **level 1** is the broadest level, subdivided into terrestrial, freshwater and marine; **level 2** has nine broad subdivisions: grassland, woodland & forest, heathland & scrub, wetland, sparsely-vegetated land, cropland, urban, rivers & lakes and marine inlets & transitional waters; **level 3** has 20 further subdivisions; **level 4** has 84 further subdivisions, mostly corresponding to UK BAP Priority Habitats; **level 5** has 152 further subdivisions, mostly based on EU Annex 1 habitats. At any level the category type, with an alpha-numeric *primary code*, can be combined with one or more of 268 numbered *secondary codes* for additional information such as presence of grazing animals, plantation woodland, ancient woodland, etc. UKHab is commonly used for surveys of sites of proposed developments, often in conjunction with the Statutory Biodiversity Metric (see below). The current (2025) version (since 2023) is UKHab v2. Further UKHab details at https://www.ukhab.org/

Irish Vegetation Classification (IVC)

A recent classification of Irish vegetation types made available between 2015 and 2021. Like the British NVC it is based on analysis of data from sample plots. It has four levels: **division** (14 categories - bogs, dunes, fens/mires/springs, freshwater habitats, grasslands, heaths, invasive plant communities, rocky habitats, scrub/wayside, saltmarsh, strandline/shingle, saline water habitats, weeds/other annuals, woodlands/forests), **group** (42 categories), **community** (186) and **sub-community** (69). Many communities do not have sub-communities. Further information, including descriptions, floristic tables, photos and distribution maps for each community, is available at https://biodiversityireland.ie/projects/ivc-classification-explorer/

Fossit Habitat Classification

An earlier classification of all habitat types in Ireland, structured with three levels: level 1 has 11 categories, level 2 has 30 categories and level 3 has 117 categories. Described in detail in *A Guide to Habitats in Ireland* by J. Fossitt, published by The Heritage Council in 2000; ISBN 1 901137 27 9.

Related systems of assessment of habitat importance/quality

Common Standards Monitoring (CSM) is a system to assess and monitor the condition of habitats from a biodiversity/conservation viewpoint. It is used by UK conservation agencies at sites with conservation designations. Different sets of criteria are used in different habitats. Details at www.jncc.gov.uk. Adaptations of CSM are used by other organisations for assessment/monitoring of other sites.

Biodiversity Metrics. The **Statutory Biodiversity Metric** provides a numerical summary/assessment of biodiversity interest (based on habitat data) of sites pre- and (estimated) post- development, to help people assess the acceptability of proposed developments in England. It includes a classification of over 130 habitat categories, many based on modified UKHab categories (see above). Further information at: https://www.gov.uk/government/publications/statutory-biodiversity-metric-tools-and-guides. A **Scottish Biodiversity Metric** is being developed; draft tool and guidance expected to be available in 2026 (https://www.nature.scot/doc/biodiversity-metric-scotlands-planning-system).

Groundwater Dependent Terrestrial Ecosystem (GWDTE) habitats. *Guidance on Assessing the Impacts of Developments on Groundwater Dependent Terrestrial Ecosystems* by the Scottish Environmental Protection Agency (SEPA) (August 2024) lists the following NVC communities as potential Groundwater Dependent Terrestrial Ecosystems (GWDTE) in Scotland: **M4-16, M21-24, M26-38, S2, S3, S7, S11, S24, S25, MG4, MG8-11, W1-7, W20, CG10-11 (not on limestone), CG12, U5c, U6, U15-17 and SD13-17**. See also Annex 1 (NVC/GWDTE table) in the UK-wide *Technical report on groundwater dependent terrestrial ecosystem (GWDTE) threshold values, v8* (UKTAG, 2012) and, for Ireland, chapter 5 in *A Framework for the Assessment of Groundwater - Dependent Terrestrial Ecosystems under the Water Framework Directive*; Environmental Research Centre Report 12; Environmental Protection Agency; 2008.

UK Biodiversity Action Plan Priority Habitats, Scottish Biodiversity List Priority Habitats and **EU Annex 1 habitats:** see pages 382-383.

FURTHER INFORMATION

Guides to plant species identification, species distributions and habitat classification:

- *Collins Wild Flower Guide, 2nd Ed.*, by D. Streeter (Harper Collins, 2016): good text; colour illustrations of most vascular plants growing wild in Britain & Ireland; ISBN 978-0-00-815675-6.
- *The Wild Flower Key*, by Francis Rose, revised & updated by Clare O'Reilly (F. Warne, 2006): detailed text; colour illustrations; omits grasses, sedges, rushes & ferns. ISBN 9780723251750.
- *Wild Flowers of Britain & Ireland, 2nd Ed.*, by M. Blamey, R. Fitter & A. Fitter (Bloomsbury Natural History 2013): text, colour illustrations and distribution maps for all vascular plants growing wild in Britain & Ireland. ISBN 978-1408179505.
- *New Flora of the British Isles (4th edition)* by C. Stace, published by C&M Floristics in 2019: covers over 5000 species/subspecies/hybrids in 1266 pages; detailed text (much technical terminology); most species are not illustrated. ISBN 9781527226302.
- *Grasses of the British Isles*, by T. Cope & A. Gray (Botanical Society of the British Isles, 2009): comprehensive, detailed and well illustrated; ISBN 9780901158420.
- *Grasses (3rd edition)*, by C.E. Hubbard (Penguin, 1982): comprehensive, very detailed and well-illustrated; still an invaluable reference; ISBN 9780140132274.
- *Sedges of the British Isles (3rd edition)*, by A.C. Jermy, D.A. Simpson, A.J.Y. Foley & M.S. Porter (Botanical Society of the British Isles, 2007) is the equivalent for sedges. ISBN 978-0-901158-35-2.
- *Ferns, clubmosses, quillworts & horsetails of Britain & Ireland* (J. Merryweather; Princeton University Press, 2020); comprehensive, detailed and well-illustrated with photos; ISBN 978-0-691-18039-7.
- *The Vegetative Key to the British Flora, 2nd edition,* by J. Poland & E. Clement (Botanical Society of the British Isles, 2009). Detailed and comprehensive. Size: A5. ISBN 978-0-9560144-2-9.
- *British mosses and liverworts: a field guide*, edited by I. Atherton, S. Bosanquet & M. Lawley (British Bryological Society, 2010): good identification guide to 75% of our 1000+ species; brief notes on the remaining 25%; ISBN 9780956131010; also online at www.britishbryologicalsociety.org.uk
- *Lichens - An Illustrated Guide to the British and Irish Species, 7th edition*; F.S. Dobson (published by The British Lichen Society / The Richmond Publishing Co. Ltd., 2018); a very useful book describing over 1000 species and including photos and distribution maps. ISBN 978-0-9565291-0-7.
- *An Illustrated Guide to British Upland Vegetation*, by A.M. Averis, A.B.G. Averis, H.J.B. Birks, D. Horsfield, D.B.A. Thompson & M.J.M. Yeo (UK Joint Nature Conservation Committee, 2004; reprinted by Pelagic Publishing, 2014) describes plant communities of N & W Britain using the NVC as a framework; ISBN 9781784270155; also available as a downloadable pdf from www.jncc.gov.uk
- The National Biodiversity Network's Gateway (NBN) (www.nbnatlas.org) is a website providing distribution maps and details of individual records of species (of all groups) in Britain and Ireland.
- *Plant Atlas 2020*, published by BSBI: distribution maps and information about habitats, status, etc. An excellent resource. ISBN 9780691247595; also online at https://plantatlas2020.org/atlas/

Organisations and other online resources concerned with botanical studies in Britain and Ireland:

- The **Botanical Society of the British Isles (BSBI)** (www.bsbi.org) is an organisation concerned mainly with the identification and recording of vascular plants in Britain and Ireland.
- The **British Bryological Society (BBS)** (www.britishbryologicalsociety.org.uk) (changing to **British and Irish Bryological Society BIBS** in 2026) is the equivalent for bryophytes (mosses and liverworts).
- The **British Lichen Society (BLS)** (www.britishlichensociety.org.uk) is the equivalent for lichens.
- **Plantlife** (www.plantlife.org.uk) is a charity concerned with the conservation of plants, including mosses, liverworts and lichens, in Britain and elsewhere in the world.
- The **National Plant Monitoring Scheme (NPMS)** is a great citizen science project run by the BSBI, Plantlife, Joint Nature Conservation Committee (JNCC) & UK Centre for Ecology & Hydrology (UKCEH) monitoring botanical interest throughout the UK since 2015. See https://www.npms.org.uk/
- The **Field Studies Council** (www.field-studies-council.org) runs courses on various aspects of plant species identification and habitat classification.
- **iSpot** (www.ispotnature.org) is a website to help people identify species of any group (plants, birds etc). People can send in photos and others can then look at them and help identify the species.

INDEX

Drawings of the main parts of a plant

The drawings below show the main aerial (above ground) parts of a plant. There are separate drawings for grasses (left), sedges (right) and a typical herb (centre). Three drawings are of particular (named) species; the other three are composites, each with features of more than one species.

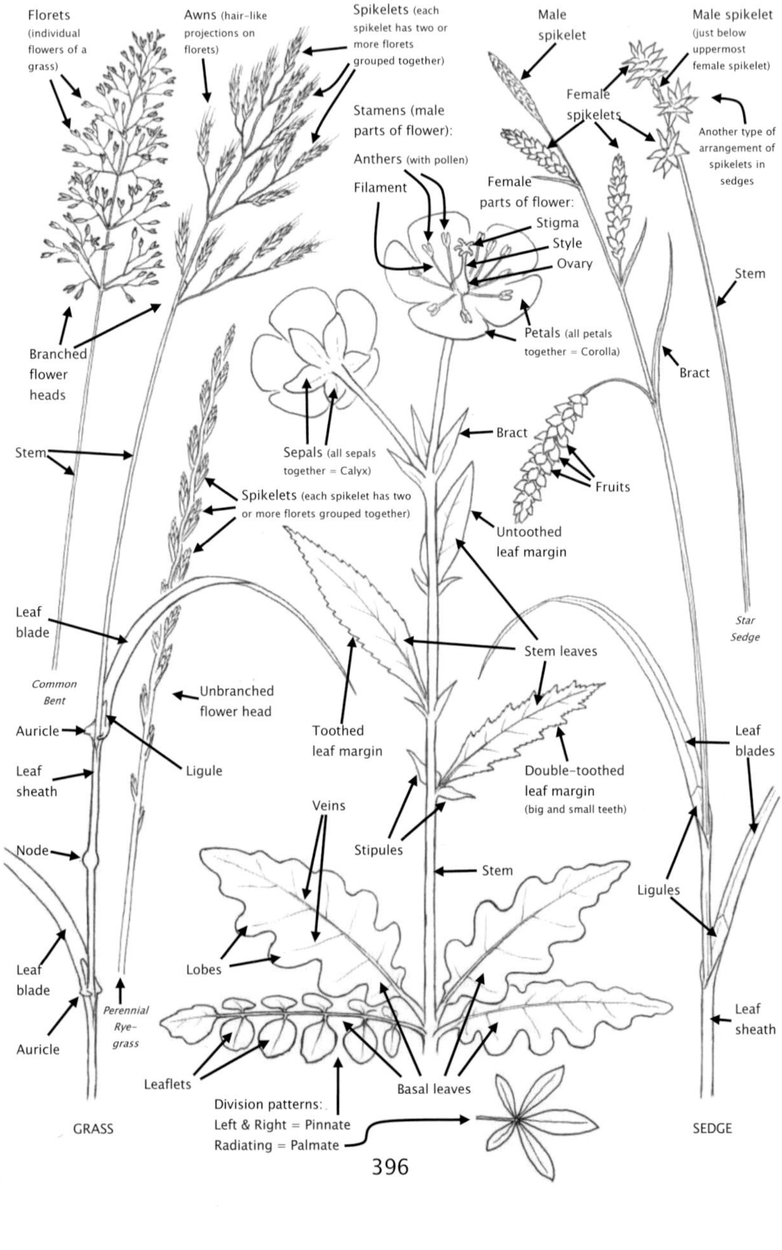